Getting Started with Raspberry Pi

Matt Richardson and Shawn Wallace

THIRD EDITION

SAN FRANCISCO, CA

Getting Started with Raspberry Pi

by Matt Richardson and Shawn Wallace

Published by Maker Media, Inc., 1160 Battery Street East, Suite 125, San Francisco, CA 94111.

Maker Media books may be purchased for educational, business, or sales promotional use. Online editions are also available for most titles (*http://safaribooksonline.com*). For more information, contact our corporate/institutional sales department: 800-998-9938 or *corporate@oreilly.com*.

Editor: Patrick Di Justo
Production Editor: Colleen Cole
Copyeditor: James Fraleigh
Proofreader: Amanda Kersey
Indexer: WordCo Indexing Services, Inc.
Interior Designer: David Futato
Cover Designer: Karen Montgomery
Illustrator: Rebecca Demarest

December 2012:	First Edition
October 2014:	Second Edition
July 2016:	Third Edition

Revision History for the Third Edition

2016-07-05   First Release

See *http://oreilly.com/catalog/errata.csp?isbn=9781680452464* for release details.

978-1-680-45246-4

[LSI]

Contents

Preface

It's easy to understand why people were skeptical of the Raspberry Pi when it was first announced in 2011. A credit-card–sized computer for \$35 seemed like a pipe dream. Which is why, when it started shipping, the Raspberry Pi created a frenzy of excitement.

Demand outstripped supply for months, and the waitlists for these minicomputers were very long. Besides the price, what is it about the Raspberry Pi that tests the patience of this hardware-hungry mass of people? Before we get into everything that makes the Raspberry Pi so great, let's talk about its intended audience.

Eben Upton and his colleagues at the University of Cambridge noticed that students applying to study computer science didn't have the skills that they did in the 1990s.

They attributed this to—among other factors—the "rise of the home PC and games console to replace the Amigas, BBC Micros, Spectrum ZX and Commodore 64 machines that people of an earlier generation learned to program on."[1]

Because the computer has become important for every member of the household, it may also discourage younger members from tinkering around and possibly putting such a critical tool out of commission for the family.

Meanwhile, mobile phone and tablet processors had become less expensive while getting more powerful, clearing the path for

[1] "About Us," Raspberry Pi Foundation (*http://www.raspberrypi.org/about*).

the Raspberry Pi's leap into the world of ultra-cheap-yet-serviceable computer boards.

As Linus Torvalds, the founder of Linux, said in an interview with BBC News, Raspberry Pi makes it possible to "afford failure."[2]

Raspberry Pi Foundation

It's important to note that Raspberry Pi primarily exists to advance the charitable mission of the Raspberry Pi Foundation. That mission is to "put the power of digital making into the hands of people all over the world." The Raspberry Pi Foundation hopes that people—kids especially—will learn to code, learn how computers work, and learn how to make things with computers.

With every Raspberry Pi purchase, you're not only paying for the cost of the hardware, fulfillment, and the engineering behind it, you're also making a contribution to the free online resources, free teacher training, and special programs that the Raspberry Pi Foundation offers to further its charitable mission.

As you'll learn in this book, the Raspberry Pi is great for learning, but it also makes a powerful tool. Even if the primary purpose of the board is for education, we find that its utilization stretches into commercial and industrial applications. Companies use it for things such as sensor networks, remote monitoring, and product prototyping. Even though the Raspberry Pi is great for kids, it's important to remember that it's a real computer. It's not a toy or some kind of watered-down device.

What Can You Do with It?

One of the great things about the Raspberry Pi is that there's no single way to use it. Whether you just want to watch videos and browse the Web, or you want to hack, learn, and make with the board, the Raspberry Pi is a flexible platform for fun, utility, and experimentation. Here are just a few of the different ways you can use a Raspberry Pi:

2 Leo Kelion, "Linus Torvalds: Linux Succeeded Thanks to Selfishness and Trust," BBC News, June 12, 2012.

General-purpose computing

It's important to remember that the Raspberry Pi is a computer and you can, in fact, use it as one. After you get it up and running in Chapter 1, you can launch a web browser to access email, news sites, and social networks, which is a lot of what we use computers for these days. Going beyond the Web, you can launch the free and open source LibreOffice (*http://www.libreoffice.org*) productivity suite, which allows you to work with documents and spreadsheets when you don't have an Internet connection.

Learning to program

Because the Raspberry Pi is meant as an educational tool to encourage kids to experiment with computers, it comes preloaded with interpreters and compilers for many different programming languages. If you're eager to jump into writing code, the Python programming language is a great way to get started, and we cover the basics of it in Chapter 4. But with Raspberry Pi, you're not limited to only Python. You can write programs for your Raspberry Pi in many different programming languages, including C, Ruby, Java, and Perl. There's even a programming language and development environment for creating music called Sonic Pi.

Project platform

The Raspberry Pi differentiates itself from a regular computer not only because of its price and size, but also because of its ability to integrate with electronics projects. Starting in Chapter 6, we'll show you how to use the Raspberry Pi to control components from LEDs to AC devices, and you'll learn how to read the state of buttons and switches.

Product prototyping

More and more electronics products use Linux computers inside, and now this world of *embedded Linux* is more accessible than ever. Let's say you create something with your Raspberry Pi that would make a great product for the everyday consumer. With the *Raspberry Pi Compute Module* (a smaller version of the board that we'll discuss later), it becomes possible to create a product that's powered by Raspberry Pi.

Raspberry Pi for Makers

As makers, we have a lot of choices when it comes to platforms on which to build technology-based projects. Lately, microcontroller development boards like the Arduino have been a popular choice because they've become very easy to work with. But *system on a chip* platforms like the Raspberry Pi are a lot different than traditional microcontrollers in many ways. In fact, the Raspberry Pi has more in common with your computer than it does with an Arduino.

This is not to say that a Raspberry Pi is better than a traditional microcontroller; it's just different. For instance, if you want to make a basic thermostat, you're probably better off using an Arduino Uno or similar microcontroller for purposes of simplicity. But if you want to be able to remotely access the thermostat via the Web to change its settings and download temperature log files, you should consider using the Raspberry Pi.

Choosing between one or the other will depend on your project's requirements, and in fact, you don't necessarily have to choose between the two. In Chapter 5, we'll show you how to use the Raspberry Pi to program the Arduino and get them communicating with each other.

As you read this book, you'll gain a better understanding of the strengths of the Raspberry Pi and how it can become another useful tool in the maker's toolbox.

But Wait...There's More!

There's so much you can do with the Raspberry Pi, we couldn't fit it all into one book. Here are several other ways you can use this computer:

Media center
> Because the Raspberry Pi has both HDMI and composite video outputs, it's easy to connect to televisions. It also has enough processing power to play fullscreen video in high definition. To leverage these capabilities, you can install set-top media player operating systems like OpenELEC (*http://openelec.tv*) and OSMC (*https://osmc.tv*) on Raspberry Pi. These systems can play many different media formats, and their interfaces are designed with large buttons and text so that they can be easily controlled from the couch. They make the Raspberry Pi a fully customizable home entertainment center component.

"Bare metal" computer hacking
> Most people who write computer programs write code that runs within an operating system, such as Windows, Mac OS, or—in the case of Raspberry Pi—Linux. But what if you could write code that runs directly on the processor without the need for an operating system? You could even write your own operating system from scratch if you were so inclined. The University of Cambridge's Computer Laboratory has published a free online course (*http://bit.ly/1BW2e3C*) which walks you through the process of writing your own OS using assembly code.

Retro gaming
> There's a huge community of retro gaming enthusiasts that use Raspberry Pi loaded up with RetroPie (*http://blog.petrockblock.com/retropie*) as a platform for emulating their old gaming systems such as Nintendo, Game Boy, Atari, and DOS. There are even a few Raspberry Pi add-on boards which make it easy to wire up your own arcade-style buttons as input. With this hardware and software combination, you could use Raspberry Pi to refurbish an old arcade cabinet and give it new life.

Linux and Raspberry Pi

Your typical computer is running an operating system, such as Windows, OS X, or Linux. It's what starts up when you turn your computer on, and it provides your applications access to hardware functions of your computer. For instance, if you're writing an application that accesses the Internet, you can use the operating system's functions to do so. You don't need to understand and write code for every single type of Ethernet or WiFi hardware out there.

Like any other computer, the Raspberry Pi also uses an operating system, and the "stock" OS is a flavor of Linux called *Raspbian*. Linux is a great match for Raspberry Pi because it's free and open source. On one hand, Linux keeps the price of the platform low, and on the other, it makes the Raspberry Pi more hackable.

And you're not limited to just Raspbian, as there are many different flavors, or *distributions*, of Linux that you can load onto the Raspberry Pi. There are even several non-Linux OS options available out there. Take a look at Chapter 3 for a rundown of different Linux and non-Linux operating systems. While creating this book, we used the standard Raspbian distribution that's available from Raspberry Pi's download page (*http://www.rasp berrypi.org/downloads*). It's a good place to start.

If you're not familiar with Linux, don't worry, Chapter 2 will equip you with the fundamentals you'll need to know to get around.

What Others Have Done with Raspberry Pi

When you have access to an exciting new technology, it can be tough deciding what to do with it. Fortunately, there's no shortage of interesting and creative Raspberry Pi projects out there to get inspiration from. At Make:, we've seen a lot of fantastic uses of the Raspberry Pi come our way, and we want to share some of our favorites:

Arcade Game Coffee Table (http://bit.ly/1oTTkxK)
Instructables user grahamgelding uploaded a step-by-step tutorial on how to make a coffee table that doubles as a classic arcade game emulator using the Raspberry Pi. To get the games running on the Pi, he used MAME (Multiple Arcade Machine Emulator), a free, open source software project that lets you run classic arcade games on modern computers. MAME is included with RetroPie, mentioned in "But Wait... There's More!" on page 11. Within the table itself, he mounted a 24-inch LCD screen connected to the Raspberry Pi via HDMI, classic arcade buttons, and a joystick connected to the Pi's general-purpose input/output (GPIO) pins to be used as inputs.

RasPod (https://github.com/lionaneesh/RasPod)
Aneesh Dogra, a teenager in India, was one of the runners-up in Raspberry Pi Foundation's 2012 Summer Coding Contest. He created Raspod, a Raspberry Pi–based, web-controlled MP3 audio player. Built with Python and a web framework called Tornado, Raspod lets you remotely log in to your Raspberry Pi to start and stop the music, change the volume, select songs, and make playlists. The music comes out of the Raspberry Pi's audio jack, so you can use it with a pair of computer speakers, or you can connect it to a stereo system to enjoy the tunes.

Raspberry Pi Supercomputer (http://bit.ly/191p9SP)
Many supercomputers are made of clusters of standard computers linked together, and computational jobs are divided among all the different processors. A group of computational engineers at the University of Southampton in the United Kingdom linked 64 Raspberry Pis to create an inexpensive supercomputer. While it's nowhere near the computational power of the top-performing supercomputers of today, it demonstrates the principles behind engineering such systems. Best of all, the rack system used to hold all these Raspberry Pis was built with Lego bricks by the team leader's six-year-old son.

If you do something interesting with your Raspberry Pi, we'd love to hear about it. You can submit your projects to the Make: editorial team through the contribute form on Makezine.com (*http://makezine.com/contribute*). You can also send us a tweet at @MattRichardson (*https://twitter.com/MattRichardson*) and @fluxly (*https://twitter.com/fluxly*).

Conventions Used in This Book

The following typographical conventions are used in this book:

Italic
> Indicates new terms, URLs, email addresses, filenames, and file extensions.

`Constant width`
> Used for program listings, as well as within paragraphs to refer to program elements such as variable or function names, databases, data types, environment variables, statements, and keywords.

`Constant width bold`
> Shows commands or other text that should be typed literally by the user.

`Constant width italic`
> Shows text that should be replaced with user-supplied values or by values determined by context.

 This element signifies a tip or suggestion.

This element signifies a general note.

 This element indicates a warning or caution.

Safari® Books Online

Safari Books Online is an on-demand digital library that delivers expert content in both book and video form from the world's leading authors in technology and business.

Technology professionals, software developers, web designers, and business and creative professionals use Safari Books Online as their primary resource for research, problem solving, learning, and certification training.

Safari Books Online offers a range of plans and pricing for enterprise, government, education, and individuals.

Members have access to thousands of books, training videos, and prepublication manuscripts in one fully searchable database from publishers like Maker Media, O'Reilly Media, Prentice Hall Professional, Addison-Wesley Professional, Microsoft Press, Sams, Que, Peachpit Press, Focal Press, Cisco Press, John Wiley & Sons, Syngress, Morgan Kaufmann, IBM Redbooks, Packt, Adobe Press, FT Press, Apress, Manning, New Riders, McGraw-Hill, Jones & Bartlett, Course Technology, and hundreds more. For more information about Safari Books Online, please visit us online.

How to Contact Us

Please address comments and questions concerning this book to the publisher:

> Make:
> 1160 Battery Street East, Suite 125
> San Francisco, CA 94111
> 877-306-6253 (in the United States or Canada)
> 707-639-1355 (international or local)

Make: unites, inspires, informs, and entertains a growing community of resourceful people who undertake amazing projects in their backyards, basements, and garages. Make: celebrates your right to tweak, hack, and bend any technology to your will. The Make: audience continues to be a growing culture and community that believes in bettering ourselves, our environment, our educational system—our entire world. This is much more than an audience, it's a worldwide movement that Make: is leading— we call it the Maker Movement.

For more information about Make:, visit us online:

Make: magazine: *http://makezine.com/magazine*
Maker Faire: *http://makerfaire.com*
Makezine.com: *http://makezine.com*
Maker Shed: *http://makershed.com*

We have a web page for this book, where we list errata, examples, and any additional information. You can access this page at *http://bit.ly/gs_with_raspberry_pi3*.

To comment or ask technical questions about this book, send email to *bookquestions@oreilly.com*.

Acknowledgments

We'd like to thank a few people who have provided their knowledge, support, advice, and feedback to *Getting Started with Raspberry Pi*:

Patrick DiJusto
Brian Jepson
Frank Teng
Anna Kaziunas France
Marc de Vinck
Eben Upton
Tom Igoe
Clay Shirky
John Schimmel
Phillip Torrone
Limor Fried
Kevin Townsend
Ali Sajjadi
Andrew Rossi

1/Getting Up and Running

A few words come up over and over when people talk about the Raspberry Pi: small, cheap, hackable, education oriented. However, it would be a mistake to describe it as *plug and play*, even though it is easy enough to plug it into a TV set and get something to appear on the screen. This is not a consumer device, and depending on what you intend to do with your Raspberry Pi, you'll need to make a number of decisions about peripherals and software when getting up and running.

Of course, the first step is to actually acquire a Raspberry Pi. Chances are you have one by now, but if not, the Raspberry Pi Foundation has arrangements with a few manufacturers from whom you can buy a Pi directly at the well-known $25–$35 price range. The official distributors are:

Premier Farnell/Element 14 (http://bit.ly/1oTTuVK)
> A British electronics distributor with many subsidiaries all over the world (such as Newark and MCM in the United States)

RS Components (http://www.rs-components.com/raspberrypi)
Another UK-based global electronics distributor (and parent of Allied Electronics in the United States)

The Raspberry Pi's low price is obviously an important part of the story. Enabling the general public to go directly to a distributor and order small quantities for the same price offered to resellers is an unusual arrangement. A lot of potential resellers were confounded by the original announcements of the price point; it was hard to see how there could be any profit margin. That's why you'll see some "downstream" resellers adding a slight markup to the $35 price (usually to $40 or so). Though the general public can still buy direct from the distributors mentioned here for the original price, the retailers and resellers often can fulfill orders faster and provide many well-curated accessories for Raspberry Pi. Here are a few of our favorite resellers:

Adafruit (https://www.adafruit.com)
Based in New York City, Limor Fried ("Ladyada") and her team have created one of the go-to ecommerce sites for things related to making. They make and sell a lot of neat Raspberry Pi accessories.

Maker Shed (http://www.makershed.com)
The official store of Make. They make some neat Raspberry Pi kits that can be found in retailers like Barnes & Noble.

Sparkfun (https://www.sparkfun.com)
Sparkfun sells lots of electronics prototyping tools and supplies, including a bunch of accessories and kits for Raspberry Pi.

Micro Center (http://www.microcenter.com)
With 25 bricks-and-mortar stores across the United States, Micro Center has positioned itself as one of the premier resellers of Raspberry Pi and tons of accessories. Browsing the aisles at Micro Center is a great way to discover new Raspberry Pi–related products.

Pimoroni (https://shop.pimoroni.com)
The gang from Pimoroni have been creating and selling fantastic Raspberry Pi products from Sheffield, UK. Fun fact:

one of their founders, Paul, is the designer of the Raspberry Pi logo!

The Pi Hut (https://thepihut.com)
The Pi Hut is another UK-based Raspberry Pi reseller. It creates many Raspberry Pi kits and also is an official distributor for Raspberry Pi Zero.

Enough microeconomic gossip; let's start by taking a closer look at the Raspberry Pi board.

A Tour of the Boards

There have been quite a few different versions of the Raspberry Pi board. The first version was the Raspberry Pi 1 Model B, which was followed by a simpler and cheaper Model A. In 2014, the Raspberry Pi Foundation announced a significant revision (and improvement) in the board design: the Raspberry Pi 1 Model B+. The Model B+ set the form-factor for "mainline" Raspberry Pis for the foreseeable future. Since then, the Foundation has also created a device for embedding the Pi in products, called the Compute Module. In 2015, it also released a stripped-down $5 model called Raspberry Pi Zero. And as of February 2016, Raspberry Pi 3 Model B is the latest mainline Raspberry Pi. A few different types of Raspberry Pis are pictured in Figure 1-1.

Over the years, there have been a few different versions of the mainline Raspberry Pi, which is the $35 model with four USB ports that most people tend to use. The versions are called Raspberry Pi 1 Model B+, Raspberry Pi 2 Model B, and Raspberry Pi 3 Model B. Each of these models added performance improvements to the processor. Raspberry Pi 2 added more RAM, and Raspberry Pi 3 added on-board WiFi and Bluetooth.

If you're following along with the examples in this book, any of these mainline Raspberry Pis will do just fine.

Figure 1-1. *There are many versions of the Pi, and here are a few (clockwise from top left): Raspberry Pi 3 Model B, Raspberry Pi 1 Model A+, Raspberry Pi Compute Module, and Raspberry Pi Zero. As of February 2016, Raspberry Pi 3 Model B is the latest mainline Raspberry Pi.*

Let's start with a tour of what you'll see when you take your Raspberry Pi out of the box.

It's tempting to think of Raspberry Pi as a microcontroller development board like Arduino or as a laptop replacement. In fact, it is more like the exposed innards of a mobile device with maker-friendly headers for various ports and functions. Figure 1-2 shows the parts of the board.

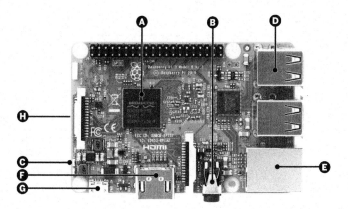

Figure 1-2. *A map of the hardware interface of the Raspberry Pi*

Here's a description of each part:

A. *The processor*

At the heart of the Raspberry Pi is the same kind of processor you'd find in a cell phone. If you're using Raspberry Pi 3, this is a 64-bit, quad core 1.2 GHz system on a chip, which is built on the ARM architecture. ARM chips come in a variety of architectures with different *cores* configured to provide different capabilities at different price points. Raspberry Pi 1 has 512 megabytes of RAM and Raspberry Pi 2 and 3 have one gigabyte of RAM.

B. *Composite video and analog audio out*

Analog audio and video outputs are available on a standard 3.5mm 4-pole plug connector. One side of the connecting cable is a four-connector mini jack (which looks like a headphone jack), and the other side is three RCA plugs for stereo audio (red and white) and composite NTSC or PAL video (yellow).

C. *Status LEDs*

Two indicator LEDs on the board provide visual feedback (Table 1-1). There are also network activity LEDs on the Ethernet jack itself.

Table 1-1. *The status LEDs*

ACT	Green	Lights when the SD card is accessed
PWR	Red	Hooked up to 3.3V power

Starting with Raspberry Pi 3, the status LEDs are placed near the MicroUSB power port as shown in Figure 1-2. For previous boards, you'll find them near the GPIO pins, in place of the WiFi antenna

D. *External USB ports*

On the mainline versions of Raspberry Pi, there are four USB 2.0 ports for connecting peripherals like keyboards, mice, thumb drives, and printers. While many USB devices can be powered from these ports, you may want to consider using a powered external hub if you have peripherals that need more power, such as a hard drive.

E. *Ethernet port*

This is a standard RJ45 Ethernet port capable of 10 or 100 megabits per second data speeds. Connect this to your router to get online; otherwise you can use WiFi.

F. *HDMI connector*

The HDMI port provides digital video and audio output. Fourteen different video resolutions are supported, and the HDMI signal can be converted to DVI (used by many monitors), composite (analog video signal usually carried over a yellow RCA connector), or SCART (a European standard for connecting audiovisual equipment) with external adapters.

G. *Power input*

One of the first things you'll realize is that there is no power switch on the Pi. This MicroUSB connector is used to supply power (this isn't an additional USB port; it's only for power). MicroUSB was selected because the connector is cheap and USB power supplies are easy to find.

H. *The microSD card slot*

You'll notice there's no hard drive on the Pi; everything is stored on a microSD card. Raspberry Pi 1 and 2 are equipped with spring-loaded slots, so you'll push to put the microSD card in and push again to take it out. With Raspberry Pi 3, they did away with the spring-loaded component in favor of a friction-fit slot. On that model, you'll push to insert the microSD card and pull to remove it. Of course, you should only do this when the Raspberry Pi is powered down.

Figure 1-3 shows all of the power and input/output (I/O) pins on the Raspberry Pi.

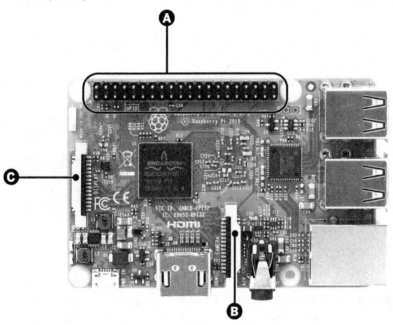

Figure 1-3. *The pins and headers on the Raspberry Pi*

Here's a description of the pins and headers shown:

A. *General-purpose input/output (GPIO) and other pins*

The current Raspberry Pis have a 2 × 20 pin GPIO header. Chapters 6 and 7 show how to use these pins to read buttons and switches and control actuators like LEDs, relays, or motors.

B. *The Camera Serial Interface (CSI) connector*

This port allows a camera module to be connected directly to the board (see Figure 1-4).

C. *The Display Serial Interface (DSI) connector*

This connector accepts a 15-pin, flat ribbon cable that can be used to communicate with the official Raspberry Pi touch display.

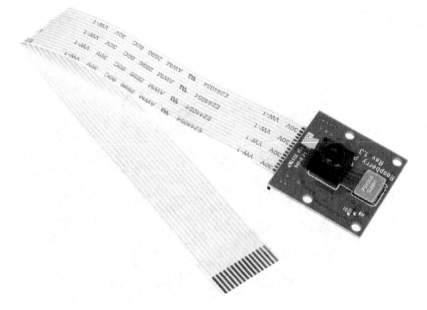

Figure 1-4. *The Raspberry Pi camera module connects directly to the CSI connector. See Chapter 9 for a full discussion of cameras and the Pi.*

The Proper Peripherals

Now that you know where everything is on the board, you'll need to know a few things about the proper peripherals to use with the Pi. There are a bunch of prepackaged starter kits that have well-vetted parts lists; there are a few caveats and gotchas when fitting out your Raspberry Pi. There's an extensive list of supported peripherals (*http://elinux.org/RPi_VerifiedPeripherals*) on the eLinux.org wiki, but these are the most basic:

A power supply

This is the most important peripheral to get right; you should use a MicroUSB adapter that can provide 5V and at least 1,500mA (1.5A) of current for Raspberry Pi 3 and 1,000 mA (1A) for Raspberry Pi 2 and older. A cell phone charger won't necessarily cut it, even if it has the correct connector. Many cell phone chargers don't provide enough current, so check the rating marked on the back. An underpowered Pi may still seem to work but will be flaky and may fail unpredictably. If in doubt, use the official Raspberry Pi power supply, which is available at most places where Raspberry Pis are sold.

There are also a number of battery-pack solutions for taking your Raspberry Pi on the go; the same rules about voltage and current apply there as well.

With the current version of the board, it is possible to power the Pi from a USB hub that feeds power back into one of the two external USB ports. However, there isn't much protection circuitry, so it may not be the best idea to power it over the external USB ports. This is especially true if you're going to be doing electronics prototyping where you may accidentally create shorts that may draw a lot of current.

A microSD card
> You'll need at least 8 GB, and it should be a Class 10 card for the best read and write performance. There are operating systems that fit onto SD cards with less than 8 GB, but the standard Raspbian installation requires at least an 8 GB microSD card.

USB keyboard and mouse
> They'll be helpful for controlling your computer. These peripherals are fairly generic, so no need to use anything fancy.

An HDMI cable
> If you're connecting to a monitor, you'll need this, or an appropriate adapter for a DVI monitor. You can also run the Pi headless, as described later in this chapter. HDMI cables can vary wildly in price. If you're just running a cable three to six feet to a monitor, there's no need to spend more than $3 on an HDMI cable. If you are running long lengths, you should definitely research the higher-quality cables and avoid the cheap generics.

Ethernet cable
> Your home may not have as many wired Ethernet jacks as it did five years ago. Because everything is wireless these days, you might find the wired port to be a bit of a hurdle; see the section "Getting Online" on page 21 for some alternatives to plugging the Ethernet directly into the wall or a hub.

WiFi USB dongle
> If you're using Pi 1 or 2, you may want to add a WiFi dongle for wireless Internet access. Many 802.11 WiFi USB dongles work with the Pi out of the box. WiFi uses a lot of power, so you will need to make sure you have an adequate power supply; a 2A supply or a powered USB hub is a good choice. If you are having problems with a WiFi dongle, power is almost always the problem.

You may also want to consider some of the following add-ons:

A powered USB hub
> If you want to add more than four USB devices to a mainline Raspberry Pi, you'll need a USB hub. A powered USB 2.0 hub is recommended.

Heatsink
> A heatsink is a small piece of metal, usually with fins, that creates a lot of surface area to dissipate heat efficiently. Heatsinks can be attached to chips that get hot. The Pi's chipset was designed for mobile applications, so a heatsink isn't necessary most of the time. However, as we'll see later, there are cases where you may want to run the Pi at higher speeds, or crunch numbers over an extended period, and the chip may heat up a bit. Some people have reported that the network chip can get warm as well.

Real-time clock
> You may want to add a real-time clock chip (like the DS1307) for logging or keeping time when offline.

Camera module
> A $25 Raspberry Pi camera module is available as an official peripheral. You can also use a USB webcam (more on this in Chapter 9).

LCD display
> Many LCDs can be used via a few connections on the GPIO header. Look for a TFT (thin-film transistor) display that can communicate with the Pi using the SPI (Serial Peripheral Interface) pins on the header. The Raspberry Pi Foundation also has a touch display that conencts to the DSI interface on the Raspberry Pi.

Sound cards
> You'll probably find that the built-in analog audio is inadequate for most of your projects. If you want high-quality sound output (or input) from the Pi, you'll need a sound card. Many USB sound cards also work well with the Pi; the Behringer's U-Control devices are a popular, inexpensive option.

Laptop dock

Several people have modified laptop docks intended for cell phones (like the Atrix lapdock) to work as a display/base for the Raspberry Pi. Some companies like Pi-Top create a laptop-like device specifically for Raspberry Pi.

HATs

A number of vendors and open hardware folks have released add-on daughterboards that sit on top of the Pi and connect via the GPIO header. These boards add capabilities like driving LCDs, motors, or analog sensor inputs. If you're familiar with Arduino terminology, you might call these daughterboards "shields," but the Raspberry Pi Foundation calls them HATs (Hardware Attached on Top), see Figure 1-5. The full specification is available on the Raspberry Pi Foundation's GitHub page (*https://github.com/raspberrypi/hats*).

Figure 1-5. *The Sense HAT add-on board includes an LED matrix, a suite of sensors, and a joystick input. It was designed for the Raspberry Pis that were sent to the International Space Station.*

To share just one example, the Raspberry Pi Foundation makes a HAT called the Sense HAT, which includes an RGB LED matrix; sensors for temperature, pressure, and humidity; an accelerometer; a gyroscope; and a magnetometer. It also has a five-position joystick. It's the HAT that was designed for the Raspberry Pis that were sent to the International Space Station as part of the Foundation's Astro Pi program.

The Case

You may find that you want a case for your Raspberry Pi. The stiff cables on all sides make it hard to keep flat, and some of the components like the SD card slot can be mechanically damaged even through normal use.

There are a bunch of premade cases available, but there are also a lot of case designs available to download and fabricate on a laser cutter or 3D printer. In general, avoid tabbed cases where brittle acrylic is used at right angles. The layered acrylic of the Pibow (*http://bit.ly/pibow3_case*) is a colorful option (Figure 1-6).

The Raspberry Pi Foundation also creates an official case, which uses a nice injection-molded design. It has multiple parts that can be removed to allow access to the GPIO pins and other components. (Figure 1-7).

It should probably go without saying, but it's one of those obvious mistakes you can make sometimes: make sure you don't put your Raspberry Pi on a conductive surface. Flip over the board and look at the bottom; there are a lot of components there and a lot of solder joints that can be easily shorted. Another reason why it's important to case your Pi!

Figure 1-6. *The colorful Pibow case from Pimoroni*

Figure 1-7. *With the official Raspberry Pi case, you can remove the top and sides to access the different parts of the board.*

Choose Your Distribution

The Raspberry Pi runs Linux for an operating system. Linux is technically just the kernel, but an operating system is much more than that—it's the total collection of drivers, services, and applications that makes the OS. A variety of flavors or distributions of the Linux OS have evolved over the years. Some of the most common on desktop computers are Ubuntu, Debian, Fedora, and Arch. Each has its own communities of users and is tuned for particular applications.

Because the Pi is based on a mobile device chipset, it has different software requirements than a desktop computer. The Broadcom processor has some proprietary features that require special "binary blob" device drivers and code that won't be included in any standard Linux distribution. And, while most desktop computers have gigabytes of RAM and hundreds of gigabytes of storage, the Pi is more limited in both regards. Special Linux distributions that target the Pi have been developed.

In this book, we will concentrate on the official Raspbian distribution, which is based on Debian. Note that raspbian.org is a community site, not operated by the Foundation. If you're looking for the official distribution, visit the Raspberry Pi Foundation's downloads page (*http://raspberrypi.org*). Other specialized distributions are explored in Chapter 3.

Flash the SD Card

Many vendors sell SD cards with the operating system preinstalled; for some people, this may be the best way to get started. Even if it isn't the latest release, you can easily upgrade once you get the Pi booted up and on the Internet.

The easiest way to get the OS on the microSD card is to use the NOOBS tool. Don't take offense; no one is questioning your computer acumen. NOOBS stands for New Out Of the Box Software and is a configuration tool that will help install the OS. You'll need an SD card (at least 8 GB) and reader, then follow these steps: when you boot up the Pi, you'll see a configuration screen with a number of OS options. Select Raspbian and hit the Install button; that's all there is to it!

For Advanced Users: Create Your Own Disk Image

The first thing you'll need to do is download one of the distributions from the Raspberry Pi Foundation's downloads page (*http://www.raspberrypi.org/downloads*) or one of the sites in Chapter 3. Note that you can't just drag the disk image onto the SD card; you'll need to make a bit-for-bit copy of the image. You'll need a card writer and a disk image utility; any inexpensive card writer will do. The instructions vary depending on the OS you're running. Unzip the image file (you should end up with a *.img* file), then follow the appropriate directions described in Appendix A.

Faster Downloads with BitTorrent

You'll see a note on the download site about downloading a torrent file for the most efficient way of downloading Raspbian. Torrents are decentralized way of distributing files; they can be much faster because you'll be pulling bits of the download from many other torrent clients rather than a single central server. You'll need a BitTorrent client if you choose this route.

Some popular BitTorrent clients are:

Vuze (http://www.vuze.com)
Integrated torrent search and download

Miro (http://www.getmiro.com)
Open source music and video player that also handles torrents

MLDonkey (http://mldonkey.sourceforge.net)
Windows and Linux-only filesharing tool

Transmission (http://www.transmissionbt.com)
Lightweight Mac and Linux-only client; also used in embedded systems

---

Booting Up

Follow these steps to boot up your Raspberry Pi for the first time:

1. Push the microSD card into the socket on the bottom of the board. On Raspberry Pi 1 and 2, it'll click into place. On Raspberry Pi 3, the microSD card won't click and is held in place with friction.

2. Plug in a USB keyboard and mouse.

3. Plug the HDMI output into your TV or monitor. Make sure your monitor is on and set to the correct input.

4. Last, plug in the power supply. It's a good habit to make sure everything else is hooked up *before* connecting the power.

If all goes well, you should see a bunch of startup log entries appearing on your screen. At the top, you'll see a Raspberry Pi logo, or four if you're using a quad core model (Raspberry Pi 2 or later). If things don't go well, skip ahead to "Troubleshooting" on page 23. These log messages show all of the processes that are launching as you boot up the Pi. You'll see the network interface be initialized, and you'll see all of your USB peripherals being recognized and logged. You can see these log messages after you log in by typing dmesg on the command line.

Configuring Your Pi

The very first time you boot up, you'll be presented with the Raspbian desktop environment. The first thing you'll want to do is set a few settings with the Raspberry Pi Configuration tool. To open it, click Menu→Preferences→Raspberry Pi Configuration (see Figure 1-8).

Next, in Figure 1-9, we'll show you which configuration options are essential and which you might want to come back to if you need them.

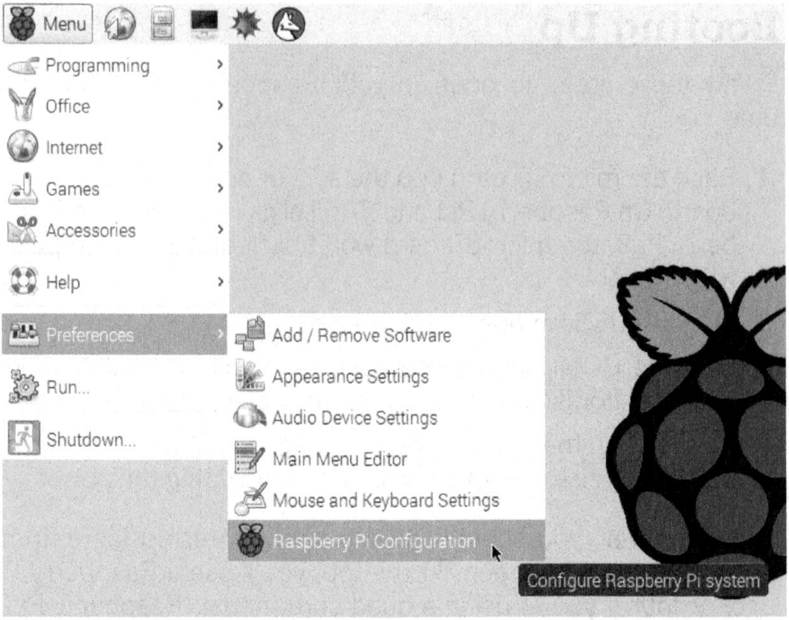

Figure 1-8. *How to launch the Raspberry Pi Configuration tool*

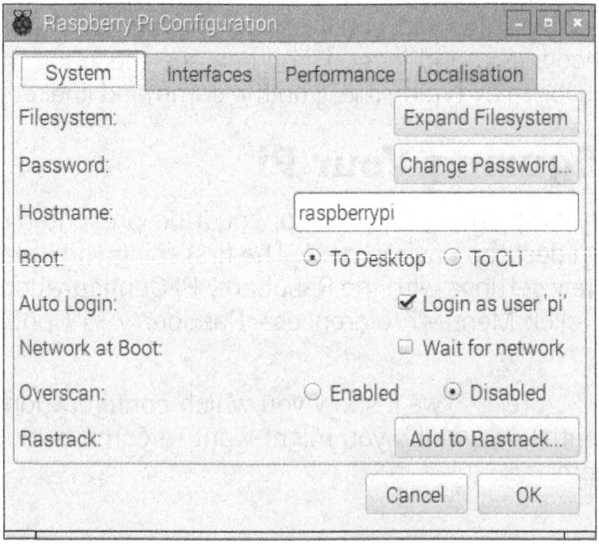

Figure 1-9. *The Raspberry Pi Configuration tool allows you to change many of the important settings on your Raspberry Pi.*

System → Expand Filesystem

You should always choose this option; this will enlarge the filesystem to let you use the whole microSD card. On the newest versions of Raspbian, this will be done automatically the first time you boot.

System → Change Password

If you're on a network with others, it's a good idea to change the default password from "raspberry" to something a little stronger.

System → Boot

This option lets you boot straight to the graphical desktop environment and is set this way by default. If you select CLI, you'll get the command line when you boot up, and you'll have to start the graphical interface manually with the command `startx`.

System → Overscan

The overscan option is set to enabled at first because some monitors may cut off the edges of the desktop. If you have a black border around your desktop, then you can disable overscan to get the desktop to fill your screen.

Interfaces → SSH

This option turns on the Secure Shell (SSH) server, which will allow you to log in to the Raspberry Pi remotely over a network. This is really handy, so you should leave it on.

Performance → GPU Memory

This option allows you to change the allocation of RAM available to the graphics processing unit. The rest of the RAM is left for the CPU to use. It's best to leave the default split for now. If you decide to experiment with 3D graphics or video decoding, you may want to adjust this value in the future.

Performance → Overclock

With this option, you can run the processor at speeds higher than the default operation. This option is not available for Raspberry Pi 3. For now, it's best to leave this setting alone.

Localisation → Set Keyboard

The default keyboard settings are for a generic keyboard in a UK-style layout. If you want the keys to do what they're labeled to do, you'll definitely want to select a keyboard type and mapping that corresponds to your setup. Luckily, the keyboard list is very robust. Note that your locale settings can affect your keyboard settings as well.

Localisation → Set Locale

If you're outside the UK, you should change your locale to reflect your language and character encoding preferences. The default setting is for UK English with a standard UTF-8 character encoding (en_GB.UTF-8). Select en_US.UTF-8 if you're in the US.

Localisation → Set Timezone

You'll probably want to set this.

When you're done, select OK and you'll be prompted to restart so that the settings can take effect.

If you want to access these settings from the command line, you can use the `raspi-config` tool (see Figure 1-10). Type the following at the command line if you want to try that out:

```
sudo raspi-config
```

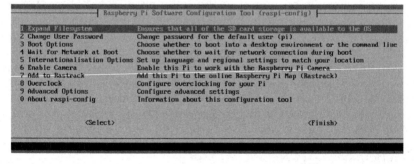

Figure 1-10. *The raspi-config tool when run from the command line*

Getting Online

You've got a few different ways to connect to the Internet. If you've got easy access to a router, switch, or Ethernet jack connected to a router, just plug in using a standard Ethernet cable. If you have a WiFi USB dongle or you're using a Raspberry Pi 3, you can connect wirelessly; there's an icon on the taskbar to set up your wireless connection (see Figure 1-11).

Figure 1-11. *Click on the network icon on the right side of the taskbar to select a WiFi network to connect to.*

If you've got a laptop nearby, or if you're running the Pi in a headless configuration, you can share the WiFi on your laptop with the Pi (Figure 1-12). It is super simple on the Mac: just enable Internet Sharing In your Sharing settings, then use an Ethernet cable to connect the Pi and your Mac. In Windows, enable "Allow other network users to connect through this computer's Internet connection" in your Internet Connection Sharing properties. The Pi should automatically get an IP address when connected and be online.

You will probably need a cross-over cable for a Windows-based PC, but you can use any Ethernet cable on Apple hardware, as it will autodetect the type of cable.

Figure 1-12. *A handy trick is to share your laptop's WiFi connection with the Pi. You can also run the Pi headless (see "Running Headless" on page 23), which is convenient if you're using your Raspberry Pi on the run.*

Shutting Down

There's no power button on the Raspberry Pi (although there is a header for a reset switch on newer boards). The proper way to shut down is through the Shutdown command under the task-bar menu within the desktop environment.

You can also shut down from the command line by typing:

```
pi@raspberrypi ~ $ sudo halt
```

Be sure to do a clean shutdown (and don't just pull the plug). In some cases, you can corrupt the SD card if you turn off power without halting the system.

Running Headless

If you want to work on the Raspberry Pi without plugging in a monitor, keyboard, and mouse, there are some ways to set it up to run *headless*. If all you require is to get into the command line, you can simply hook the Raspberry Pi up to the network and use an SSH client to connect to it (username: **pi**, password: **rasp berry**). The SSH utility on Mac or Linux will do; use PuTTY (*http://bit.ly/1sfuf4X*) on Windows (or Linux). The SSH server on the Raspberry Pi is enabled by default (run the Raspberry Pi configuration utility again if for some reason it doesn't launch at startup).

Another way to connect to the Pi over a network connection is to run a Virtual Network Computing (VNC) server on the Pi and connect to it using a VNC client. The benefit of this is that you can run a complete working graphical desktop environment in a window on your laptop or desktop. This is a great solution for a portable development environment. See the Raspberry Pi Hub (*http://elinux.org/RPi_VNC_Server*) for extensive instructions on how to install TightVNC (*http://www.tightvnc.com*), a lightweight VNC server.

A third way of logging in to the Pi without a keyboard or monitor is via some pins on the GPIO header. You can use a special cable from FTDI that allows you to connect to that serial port via USB. The FTDI cable has three wires that connect to ground (pin 6), TX (pin 8), and RX (pin 10) on the header. Alternatively, you could use the BUB I from Modern Device, which is a breakout board for the FTDI chip with a prototyping area that allows you to reroute the signals.

Troubleshooting

If things aren't working the way you think they should, there are a few common mistakes and missed steps. Be sure to check all of the following suggestions:

- Is the microSD card in the slot, and is it making a good connection? Are you using the correct type of microSD card?
- Was the disk image written correctly to the card? Try copying it again with another card reader.

- Check the integrity of your original disk image. You can do this by running a Secure Hash Algorithm (SHA) checksum utility on the disk image and comparing the result to the 40-character hash published on the download page.

- Is the Pi restarting or having intermittent problems? Check your power supply; an underpowered board may seem to work but act flaky.

- Do you get a kernel panic on startup? A kernel panic is the equivalent of Windows' Blue Screen of Death; it's most often caused by a problem with a device on the USB hub. Try unplugging USB devices and restarting.

If that all fails, head over to the Raspberry Pi Hub's trouble-shooting page (*http://elinux.org/R-Pi_Troubleshooting*) for solutions to all sorts of problems people have had.

Which Board Do You Have?

If you're asking for help in an email or on a forum, it can be helpful to those assisting you if you know exactly what version of the operating system and which board you're using. To find out the OS version, open LXTerminal and type:

```
cat /proc/version
```

To find your board version, type:

```
cat /proc/cpuinfo
```

Going Further

The Raspberry Pi Hub (http://elinux.org/RPi_Hub)
    Hosted by elinux.org, this is a massive wiki of information on the Pi's hardware and configuration.

List of Verified Peripherals (http://elinux.org/RPi_VerifiedPeripherals)
    The definitive list of peripherals known to work with the Raspberry Pi.

2/Getting Around Linux on the Raspberry Pi

If you're going to get the most out of your Raspberry Pi, you'll need to learn a little Linux. The goal of this chapter is to present a whirlwind tour of the operating system and give you enough context and commands to get around the filesystem, install packages from the command line or desktop environment, and point out the most important tools you'll need day to day.

Raspbian comes with the Lightweight X11 Desktop Environment (LXDE) graphical desktop environment installed (Figure 2-1). This is a trimmed-down desktop environment for the X Window System that has been powering the GUIs of Unix and Linux computers since the 1980s. Some of the tools you see on the desktop and in the menu are bundled with LXDE (the Leafpad text editor and the LXTerminal shell, for instance).

Running on top of LXDE is Openbox, a *window manager* that handles the look and feel of windows and menus. If you want to tweak the appearance of your desktop, click the Raspberry menu in the upper left, then choose Preferences→Appearance Settings. Unlike OS X or Windows, it is relatively easy to completely customize your desktop environment or install alternative window managers. Some of the other distributions for

Raspberry Pi have different environments tuned for applications like set-top media boxes, phone systems, or network firewalls. See *http://elinux.org/RPi_Distributions* and Chapter 3 for more.

As of October 2015's update to Raspbian, the default behavior is to log in and launch the desktop environment immediately after booting. If you find yourself at a prompt for a username and password, the default user is **pi** and the password is **raspberry**. If you find yourself at the text command line (which looks like **pi@raspberrypi ~ $**) and want to launch the desktop environment, just type **startx** and press Enter. We'll cover the command line more in-depth later in this chapter.

Figure 2-1. *The graphical desktop*

The Raspberry Pi software engineers and designers have customized Raspbian Linux and its desktop environment for general-purpose computing, making, and learning. If you browse the programs in the Raspberry menu in the upper-left corner of your screen (Figure 2-2), you'll notice that there are programming environments, office tools, accessories, games, and Internet programs preinstalled. Feel free to click around and explore!

Figure 2-2. *The contents of the Raspberry menu*

A few applications that you encounter may be a little different than those in other desktop environments:

The File Manager
   If you prefer not to move files around using the command line (more on that in a moment), select the File Manager from the taskbar or within the Raspberry menu→Accessories. You'll be able to browse the filesystem using icons and folders the way you're probably used to doing.

The web browser
   The default web browser has the code name Epiphany (*https://wiki.gnome.org/Apps/Web*), and it works well with limited resources. It's easy to forget how much work web browsers do these days. Because Raspbian is designed to be a very lightweight OS distribution, there are a number of features you may expect in a web browser that may not be available. There are a couple of other browser options, notably Chromium (*http://elinux.org/RPi_Chromium*), which is a popular choice among Raspberry Pi users. You'll have to

install it separately (see "Installing New Software" on page 42).

Video and audio

Multimedia playback is handled by `omxplayer`, which is complicated but powerful. It is only available as a command-line utility. Omxplayer is specially designed to work with the Graphics Processing Unit (GPU) on the processor; other free software like VLC and mPlayer won't work well with the GPU.

To keep the price down, certain video licenses were not included with the Raspberry Pi. If you want to watch recorded TV and DVDs encoded in the MPEG-2 format (or Microsoft's VC-1 format), you'll need to purchase a license key from the Foundation's online shop (*http://swag.raspberrypi.org*); the cost is less than \$5. A license for H.264 (MPEG-4) decoding and encoding is included with the Raspberry Pi.

Wolfram and Mathematica

Bundled with Raspbian is a pilot release of the Wolfram language and Mathematica. In fact, you'll notice icons for both in the taskbar. Mathematica is the frontend interface for the Wolfram programming language. Together, they're commonly used for complex computations in math, science, and engineering fields. To see what Wolfram and Mathematica are capable of, the Wolfram Language and System Documentation (*http://bit.ly/1qoitzW*) is a great place to start.

Text editor

Leafpad (*http://wiki.lxde.org/en/Leafpad*) is the default text editor, which is available under Raspberry menu→Accessories. For editing text files outside the desktop environment, you can use Nano (*http://www.nano-editor.org*), which is an easy-to-learn, bare-bones text editor. Vim is also installed; execute it with the command **vi**. Other common Unix text editors like Emacs are not installed by default, but can be easily added (see "Installing New Software" on page 42).

Copy and paste

Copy and paste functions work between applications pretty well, although you may find some oddball programs that aren't consistent. If your mouse has a middle button, you can select text by highlighting it as you normally would (click and drag with the left mouse button) and paste it by pressing the middle button while you have the mouse cursor over the destination window.

The shell

A lot of tasks are going to require you to get to the command line and run commands there. The LXTerminal program provides access to the command line or shell. It can be launched from the taskbar icon or from the Raspberry menu→Accessories. The default shell on Raspbian is the Bourne-again shell (bash (*http://bit.ly/1oTTXqW*)), which is very common on Linux systems. There's also an alternative called dash (*http://bit.ly/1oTTVzs*). You can change shells via the program menu or with the **chsh** command.

Using the Command Line

If it helps, you can think of using the command line as playing a text adventure game, but with the files and the filesystem in place of grues and mazes. If that metaphor doesn't help you, don't worry: all the commands and concepts in this section are standard Linux and are valuable to learn.

Before you start, open up the LXTerminal program (Figure 2-3). There are two tricks that make life much easier in the shell: *autocomplete* and *command history*. Often you will only need to type the first few characters of a command or filename, then hit Tab. The shell will attempt to autocomplete the string based on the files in the current directory or programs in commonly used directories (the shell will search for executable programs in places like */bin* or */usr/bin/*). If you hit the up arrow on the command line, you'll be able to step back through your command history, which is useful if you mistyped a character in a long string of commands.

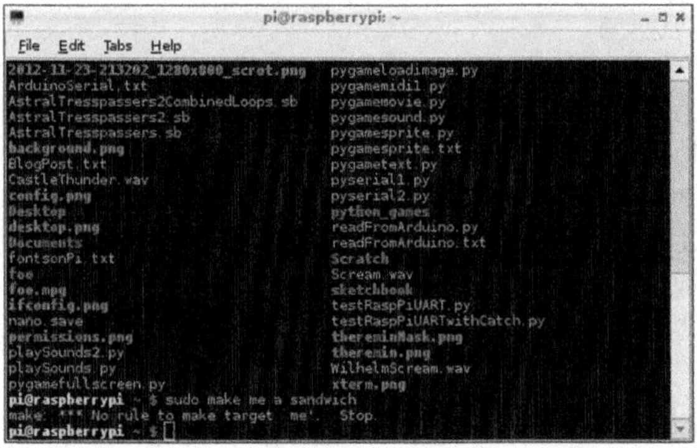

Figure 2-3. *LXTerminal gives you access to the command line (or shell)*

Files and the Filesystem

Table 2-1 shows some of the important directories in the filesystem. Most of these follow the Linux standard of where files should go; a couple are specific to the Raspberry Pi. The /sys directory is where you can access all of the hardware on the Raspberry Pi.

Table 2-1. *Important directories in the Raspbian filesystem*

Directory	Description
/bin	Programs and commands that all users can run
/boot	All the files needed at boot time
/dev	Special files that represent the devices on your system
/etc	Configuration files
/etc/init.d	Scripts to start up services
/etc/X11	X11 configuration files
/home	User home directories
/home/pi	Home directory for Pi user
/lib	Kernel modules/drivers
/media	Mount points for removable media
/proc	Virtual directory with details about running processes and the OS
/sbin	Programs for system maintenance

Directory	Description
/sys	A special directory on the Raspberry Pi that represents the hardware devices
/tmp	Space for programs to create temporary files
/usr	Programs and data usable by all users
/usr/bin	Most of the programs in the operating system reside here
/usr/games	Yes, games
/usr/lib	Libraries to support common programs
/usr/local	Software that may be specific to this machine goes here
/usr/sbin	More system administration programs
/usr/share	Supporting files that aren't specific to any processor architecture
/usr/src	Linux is open source; here's the source!
/var	System logs and spool files
/var/backups	Backup copies of all the most vital system files
/var/cache	Programs such as apt-get cache their data here
/var/log	All of the system logs and individual service logs
/var/mail	All user email is stored here, if you're set up to handle email
/var/spool	Data waiting to be processed (e.g., incoming email, print jobs)

You'll see your current directory displayed before the command prompt. In Linux, your home directory has a shorthand notation: the tilde (~). When you open the LXTerminal, you'll be dropped into your home directory, and your prompt will look like this:

```
pi@raspberrypi ~ $
```

Here's an explanation of that prompt:

```
pi@ ❶raspberrypi ❷ ~ ❸ $ ❹
```

❶ Your username, pi, followed by the at (@) symbol.

❷ The name of your computer (raspberrypi is the default hostname).

❸   The *current working directory* of the shell. You always start out in your home directory (~).

❹   This is the *shell prompt*. Any text you type will appear to the right of it. Press Enter or Return to execute each command you type.

Later in the book, we will omit the `pi@raspberrypi` ~ portion of the prompt and just show you the `$` in some examples, to keep them less cluttered.

Use the `cd` (change directory) command to move around the filesystem. The following two commands have the same effect (changing to the home directory) for the Pi user:

```
cd /home/pi/
cd ~
```

If the directory path starts with a forward slash, it will be interpreted as an absolute path to the directory. Otherwise, the directory will be considered relative to the current working directory. You can also use `.` and `..` to refer to the current directory and the current directory's parent. For example, to move up to the top of the filesystem:

```
pi@raspberrypi ~ $ cd ..
pi@raspberrypi /home $ cd ..
```

You could also get there with the absolute path `/`:

```
pi@raspberrypi ~ $ cd /
```

Once you've changed to a directory, use the `ls` command to list the files there:

```
pi@raspberrypi / $ ls
bin   dev  home  lost+found  mnt  proc  run   selinux  sys  usr
boot  etc  lib   media       opt  root  sbin  srv      tmp  var
```

Most commands have additional parameters, or *switches*, that can be used to turn on different behaviors. For example, the `-l` switch will produce a more detailed listing, showing file sizes, dates, and permissions:

```
pi@raspberrypi ~ $ ls -l
total 8
```

```
drwxr-xr-x 2 pi pi 4096 Oct 12 14:26 Desktop
drwxrwxr-x 2 pi pi 4096 Jul 20 14:07 python_games
```

The -a switch will list all files, including invisible ones:

```
pi@raspberrypi ~ $ ls -la
total 80
drwxr-xr-x 11 pi   pi   4096 Oct 12 14:26 .
drwxr-xr-x  3 root root 4096 Sep 18 07:48 ..
-rw-------  1 pi   pi     25 Sep 18 09:22 .bash_history
-rw-r--r--  1 pi   pi    220 Sep 18 07:48 .bash_logout
-rw-r--r--  1 pi   pi   3243 Sep 18 07:48 .bashrc
drwxr-xr-x  6 pi   pi   4096 Sep 19 01:19 .cache
drwxr-xr-x  9 pi   pi   4096 Oct 12 12:57 .config
drwx------  3 pi   pi   4096 Sep 18 09:24 .dbus
drwxr-xr-x  2 pi   pi   4096 Oct 12 14:26 Desktop
-rw-r--r--  1 pi   pi     36 Sep 18 09:35 .dmrc
drwx------  2 pi   pi   4096 Sep 18 09:24 .gvfs
drwxr-xr-x  2 pi   pi   4096 Oct 12 12:53 .idlerc
-rw-------  1 pi   pi     35 Sep 18 12:11 .lesshst
drwx------  3 pi   pi   4096 Sep 19 01:19 .local
-rw-r--r--  1 pi   pi    675 Sep 18 07:48 .profile
drwxrwxr-x  2 pi   pi   4096 Jul 20 14:07 python_games
drwx------  4 pi   pi   4096 Oct 12 12:57 .thumbnails
-rw-------  1 pi   pi     56 Sep 18 09:35 .Xauthority
-rw-------  1 pi   pi    300 Oct 12 12:57 .xsession-errors
-rw-------  1 pi   pi   1391 Sep 18 09:35 .xsession-errors.old
```

Use the mv command to rename a file. The **touch** command can be used to create an empty dummy file:

```
pi@raspberrypi ~ $ touch foo
pi@raspberrypi ~ $ ls
foo   Desktop   python_games
pi@raspberrypi ~ $ mv foo baz
pi@raspberrypi ~ $ ls
baz   Desktop   python_games
```

Remove a file with rm. To remove a directory, you can use rmdir if the directory is empty, or rm -r if it isn't. The -r is a parameter sent to the rm command that indicates it should recursively delete everything in the directory.

If you want to find out all the parameters for a particular command, use the man command (or you can often use the --help option):

```
pi@raspberrypi ~ $ man curl
pi@raspberrypi ~ $ rm --help
```

To create a new directory, use mkdir. To bundle all of the files in a directory into a single file, use the tar command, originally created for tape archives. You'll find a lot of bundles of files or source code are distributed as tar files, and they're usually also compressed using the gzip command. Try this:

```
pi@raspberrypi ~ $ mkdir myDir
pi@raspberrypi ~ $ cd myDir
pi@raspberrypi ~ $ touch foo bar baz
pi@raspberrypi ~ $ cd ..
pi@raspberrypi ~ $ tar -cf myDir.tar myDir
pi@raspberrypi ~ $ gzip myDir.tar
```

You'll now have a *.tar.gz* archive of that directory that can be distributed via email or the Internet.

More Linux Commands

One of the reasons that Linux (and Unix) is so successful is that the main design goal was to build a very complicated system out of small, simple modular parts that can be chained together. You'll need to know a little bit about two pieces of this puzzle: *pipes* and *redirection*.

Pipes are simply a way of chaining two programs together so the output of one can serve as the input to another. All Linux programs can read data from *standard input* (often referred to as *stdin*), write data to *standard output* (*stdout*), and throw error messages to *standard error* (*stderr*). A pipe lets you hook up stdout from one program to stdin of another (Figure 2-4). Use the | operator, as in this example:

```
pi@raspberrypi ~ $ ls -la | less
```

(Press q to exit the less program.)

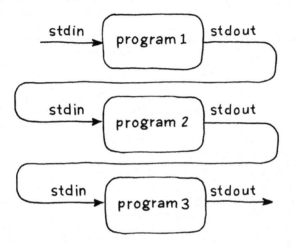

Figure 2-4. *Pipes are a way of chaining smaller programs together to accomplish bigger tasks*

Now (for something a little more out there) try:

```
pi@raspberrypi ~ $ sudo cat /boot/kernel.img | aplay
```

You may want to turn the volume down a bit first; this command reads the kernel image and spits all of the 1s and 0s at the audio player. That's what your kernel sounds like!

In some of the examples later in the book, we'll also be using *redirection*, where a command is executed and the stdout output can be sent to a file. As you'll see later, many things in Linux are treated as ordinary files (such as the Pi's general-purpose input/output pins), so redirection can be quite handy. To redirect output from a program, use the > operator:

```
pi@raspberrypi ~ $ ls > directoryListing.txt
```

Special Control Keys

In addition to the keys for autocomplete (Tab) and command history (up arrow) previously mentioned, there are a few other special control keys you'll need in the shell. Here are a few:

Ctrl-C
Kills the running program. May not work with some interactive programs such as text editors.

Ctrl-D
Exits the shell. You must type this at the command prompt by itself (don't type anything after the $ before hitting Ctrl-D).

Ctrl-A
Moves the cursor to the beginning of the line.

Ctrl-E
Moves the cursor to the end of the line.

There are others, but these are the core keyboard shortcuts you'll use every day.

Sometimes you'll want to display the contents of a file on the screen. If it's a text file and you want to read it one screen at a time, use less:

```
pi@raspberrypi ~ $ ls > flob.txt
pi@raspberrypi ~ $ less flob.txt
```

If you want to just dump the entire contents of a file to standard output, use cat (short for concatenate). This can be handy when you want to feed a file into another program or redirect it somewhere.

For example, this is the equivalent of copying one file to another with a new name (the second line concatenates the two files first):

```
pi@raspberrypi ~ $ ls > wibble.txt
pi@raspberrypi ~ $ cat wibble.txt > wobble.txt
pi@raspberrypi ~ $ cat wibble.txt wobble.txt > wubble.txt
```

To look at just the last few lines of a file (such as the most recent entry in a log file), use `tail` (to see the beginning, use `head`). If you are searching for a string in one or more files, use the venerable program `grep`:

```
pi@raspberrypi ~ $ grep Puzzle */*
```

`grep` is a powerful tool because of the rich language of *regular expressions* that was developed for it. Regular expressions can be a bit difficult to read, and may be a major factor in whatever reputation Linux has for being opaque to newcomers.

Processes

Every program on the Pi runs as a separate process; at any particular point, you'll have dozens of processes running. When you first boot up, about 75 processes will start, each one handling a different task or service. To see all these processes, run the **top** program, which will also display CPU and memory usage. **top** will show you the processes using the most resources; use the **ps** command to list all the processes and their ID numbers (see Figure 2-5).

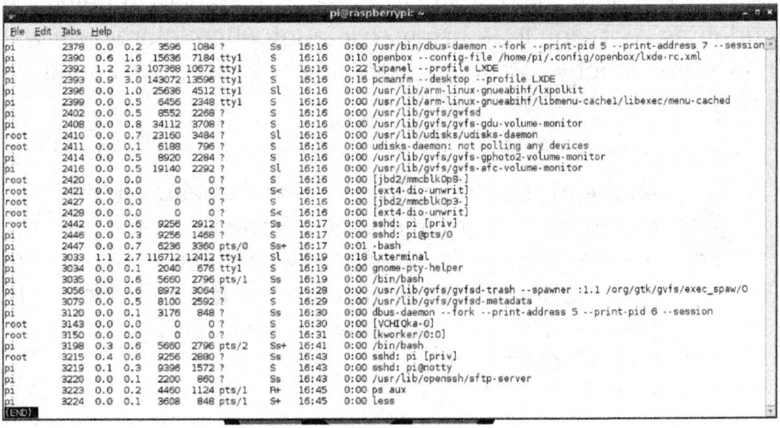

Figure 2-5. *The ps command shows a snapshot of all of the processes currently running on your Pi*

Try:

```
pi@raspberrypi ~ $ ps aux | less
```

Sometimes you may want to kill a rogue or unresponsive process. To do that, use ps to find its ID, then use kill to stop it:

```
pi@raspberrypi ~ $ kill 95689
```

In the case of some system processes, you won't have permission to kill it (though you can circumvent this with **sudo**, covered in the next section).

Sudo and Permissions

Linux is a multiuser operating system; the general rule is that everyone owns their own files and can create, modify, and delete them within their own space on the filesystem. The root (or super) user can change any file in the filesystem, which is why it is good practice to not log in as root on a day-to-day basis.

There are some tools like **sudo** ("superuser do") which allow users to act like superusers for performing tasks, like installing software without the dangers (and responsibilities) of being logged in as root. You'll be using **sudo** a lot when interacting with hardware directly, or when changing system-wide configurations.

 As the **pi** user, there's not much damage you can do to the system. As superuser, you can wreak havoc, accidentally or by design. Be careful when using sudo, especially when moving or deleting files. Of course, if things go badly, you can always make a new SD card image (see Appendix A).

Each file belongs to one user and one group. Use **chown** and **chgrp** to change the file's owner or group. You must be root to use either:

```
pi@raspberrypi ~ $ sudo chown pi garply.txt
pi@raspberrypi ~ $ sudo chgrp staff plugh.txt
```

Each file also has a set of *permissions* that show whether a file can be read, written, or executed. These permissions can be set for the owner of the file, the group, or for everyone (see Figure 2-6).

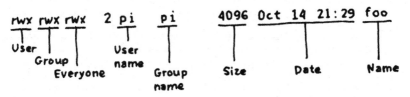

Figure 2-6. *File permissions for owner, group, and everyone*

You set the individual permissions with the chmod command. The switches for chmod are summarized in Table 2-2.

Table 2-2. *The switches that can be used with chmod*

u	User
g	Group
o	Others not in the group
a	All/everyone
r	Read permission
w	Write permission
x	Execute permission
+	Add permission
-	Remove permission

Here are a few examples of how you can combine these switches:

```
chmod u+rwx,g-rwx,o-rwx wibble.txt  ❶
chmod g+wx wobble.txt  ❷
chmod -rw,+r wubble.txt  ❸
```

❶   Allow only the user to read, write, and execute.

❷   Add permission to write and execute to entire group.

❸   Make read-only for everyone.

The only thing protecting your user space and files from other people is your password, so you better choose a strong one. Use the `passwd` command to change it, especially if you're putting your Pi on a network.

The Network

Once you're on a network, there are a number of Linux utilities that you'll be using on a regular basis. When you're trouble-shooting an Internet connection, use `ifconfig`, which displays all of your network interfaces and the IP addresses associated with them (see Figure 2-7).

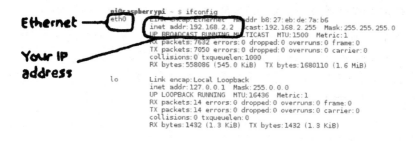

Figure 2-7. The `ifconfig` command gives you information about all of your network interfaces

The `ping` command is actually the most basic tool for trouble-shooting network connections. You can use ping (think sonar) to test whether there is a two-way connection between two IP addresses on the network or Internet. Note that many websites block ping traffic, so you may need to ping multiple sites to accurately test a connection:

```
ping yahoo.com
ping altavista.com
ping lycos.com
ping netscape.com  ❶
```

❶   Fail happens here

To log in to another computer remotely (and securely, with encrypted passwords), you can use the Secure Shell (SSH). The computer on the remote side needs to be running an SSH server for this to work, but the SSH client comes built into Raspbian. In fact, this is a great way to work on your Raspberry Pi without a monitor or keyboard, as discussed in "Running Headless" on page 23.

Related to SSH is the sftp program, which allows you to securely transfer files from one computer to another. Rounding out the set is scp, which you can use to copy files from one computer to another over a network or the Internet. The key to all of these tools is that they use the Secure Sockets Layer (SSL) to transfer files with encrypted login information. These tools are all standard stalwart Linux tools.

/etc

The /etc directory holds all of the system-wide configuration files and startup scripts. When you ran the configuration scripts the first time you started up, you were changing values in various files in the /etc directory. You'll need to invoke superuser powers with sudo to edit files in /etc; if you come across some tutorial that tells you to edit a configuration file, use a text editor to edit and launch it with sudo:

```
pi@raspberrypi ~ $ sudo nano /etc/hosts
```

Setting the Date and Time

A typical laptop or desktop will have additional hardware and a backup battery (usually a coin cell) to save the current time and date. The Raspberry Pi does not, but Raspbian is configured to automatically synchronize its time and date with a Network Time Protocol (NTP) server when plugged into a network.

Having the correct time can be important for some applications (see the example in Chapter 6 using cron to control a lamp). To set the time and date manually, use the date program:

```
$ sudo date --set="Sun Nov 18 1:55:16 EDT 2012"
```

If the time was set automatically via the Internet with NTP, you may want to update your timezone. In order to do this, go to the Internationalisation Settings within the raspi-config utility (see "Booting Up" on page 17).

Installing New Software

One of the areas where Linux completely trounces other operating systems is in software package management. *Package managers* handle the downloading and installation of software, and they automatically handle downloading and installing dependencies. Keeping with the modular approach, many software packages on Linux depend on other pieces of software. The package manager keeps it all straight, and the package managers on Linux are remarkably robust.

Raspbian comes with a pretty minimal set of software, so you will soon want to start downloading and installing new programs. The examples in this book will all use the command line for this, because it is the most flexible and quickest way of installing software.

The program `apt-get` with the `install` option is used to download software. `apt-get` will even download any extra software or libraries required so you don't have to go hunting around for dependencies. Software has to be installed with superuser permissions, so always use `sudo` (this command installs the Emacs text editor):

```
pi@raspberrypi ~ $ sudo apt-get install emacs
```

 Taking a Screenshot

One of the first things we needed to figure out when writing this book was how to take screenshots on the Pi. Pre-installed in Raspbian is a program called scrot (an abbreviation for SCReenshOT). To take a screenshot, just type `scrot` on the command line. A *.png* file will be saved in your home directory. Scrot is powerful and has a lot of command-line options; type `scrot -h` to see a short guide on how to use it. Another way to invoke scrot is to press your keyboard's *Print Screen* button.

Sound in Linux

Raspberry Pi has the built-in capability to play sound. This makes it a popular platform for DIY projects that play sound effects or stream music from the Internet. Raspbian uses the *Advanced Linux Sound Architecture*, or ALSA, for low-level control of audio devices. You can test it with a pre-loaded sound file and the sound playback utility `aplay`:

```
$ aplay /usr/share/scratch/Media/Sounds/Human/PartyNoise.wav
```

If you're using an HDMI display, by default, the sound will be played through HDMI. To force your Raspberry Pi to play through the analog output jack, run `sudo raspi-config` and choose Advanced Options→Audio→Force 3.5mm (*headphone*) jack.

To adjust the volume output, run `alsamixer` and use your arrow keys to change the gain.

You're not limited to the onboard sound; you can also add USB audio devices. Many USB audio devices even have an audio input so that you can use your Raspberry Pi to record audio in addition to playing it.

To enable a USB audio device, you'll need to edit the ALSA configuration file:

```
$ sudo nano /etc/modprobe.d/alsa-base.conf
```

Find the line that says:

```
options snd-usb-audio index=-2
```

And change it to:

```
options snd-usb-audio index=0
```

Save the file and reboot your Raspberry Pi. Now your Raspberry Pi will default to using the USB audio device.

Upgrading Your Firmware

Some of your Raspberry Pi's *firmware* is stored on the SD card and includes much of the low-level instructions that need to be executed before the boot process is handed over to your operating system.

While it's typically not necessary, if you run into some strange behavior, you may want try to update the firmware on your SD card. With an Internet-connected Raspberry Pi, this is very easy to do:

```
$ sudo rpi-update
```

If you'd like to see what's being updated when you run the utility, you can review the latest changes in Raspberry Pi's firmware repository on GitHub (*https://github.com/raspberrypi/firmware*).

To view what version of the Raspberry Pi firmware you currently have, run `vcgencmd version`:

```
$ vcgencmd version
Sep 23 2015 12:12:01
Copyright (c) 2012 Broadcom
version c156d00b148c30a3ba28ec376c9c01e95a77d6d5 (clean)
(release)
```

Going Further

There's much more to Linux and many places to continue learning about it. Some good starting points are:

Linux Pocket Guide, *by Daniel J. Barrett*
　　Handy as a quick reference.

Linux in a Nutshell, *by Ellen Siever, Stephen Figgins, Robert Love, and Arnold Robbins*
　　More detailed, but still a quick reference guide.

The Debian Wiki (*http://wiki.debian.org/FrontPage*)
　　Raspbian is based on Debian, so a lot of the info on the Debian wiki applies to Raspbian as well.

Eric S. Raymond's "The Jargon File" (*http://catb.org/jargon*)
　　Also published as the *New Hacker's Dictionary*, this collection of definitions and stories is required reading on the Unix/Linux subculture.

3/Other Operating Systems and Linux Distributions

The stock Raspbian operating system is great for general-purpose computing, but sometimes you may want to tailor the Pi to a specific purpose, like a standalone media center or guitar effects pedal. The Linux ecosystem is rich in software for every imaginable application. A number of folks have spent the time to bundle all the right software together so you don't have to do it. This chapter will highlight just a few of the more specialized Linux distributions and other operating systems to get you started.

When talking about "Linux distributions," we're usually talking about three things together:

- The Linux kernel and drivers
- Preinstalled software for a particular application
- Special configuration tools or tools preconfigured for a particular task (e.g., to boot up into a particular program)

As you saw in Chapter 1, there are essentially four popular general-purpose distributions. You'll find the first three hereafter in the NOOBS installer:

Raspbian (http://raspbian.org)
> The recommended distribution from the Foundation to start with; based on Debian. If you're not sure which distribution to choose, this is the one for you.

Arch Linux (http://www.archlinux.org)
> Arch Linux specifically targets ARM-based computers, so they supported the Pi very early on.

Pidora (http://pidora.ca)
> Pidora is a version of the Fedora distribution tuned for the Pi.

Ubuntu MATE (https://ubuntu-mate.org)
> Ubuntu MATE is a version of the very popular Ubuntu distribution of Linux. It has a slimmed-down desktop environment that works rather well on Raspberry Pi 2 and Raspberry Pi 3. It won't work on previous versions of Raspberry Pi because Ubuntu only supports ARMv7 and later.

Here are a few other interesting specialized Linux distributions and operating systems.

Distributions for Home Theater

A long-time favorite operating system for home theater is XBMC, which began as a media center project to run on the Xbox game console. Over the years, however, XBMC morphed to become a more general entertainment center platform and happens to work very well on the Pi. In the summer of 2014, the XBMC Foundation renamed the software Kodi to bring the evolution of the project into focus, because it doesn't even run on the newer Xbox versions. There are a couple of Pi distributions that make it easy to put Kodi in your living room:

OSMC (https://osmc.tv)
> Formerly called Raspbmc, the OSMC distribution is based on Debian and Kodi. It has great support for Raspberry Pi and can be installed from NOOBS. (Figure 3-1).

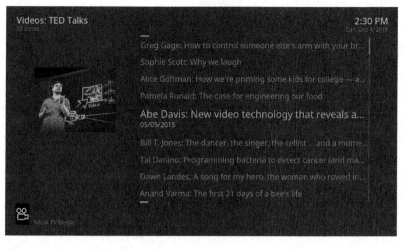

Figure 3-1. *OSMC's interface.*

OpenELEC (http://openelec.tv)

The Open Embedded Linux Entertainment Center is a pared-down version of Kodi that may appeal to more ascetic Pi users (Figure 3-2).

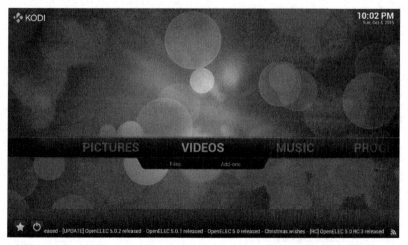

Figure 3-2. *The main menu for The Open Embedded Linux Entertainment Center (OpenELEC)*

Distributions for Music

It's cheap and it can fit in a guitar effects stomp box, so of course the electronic music world has been excited about the Pi since its release. Here are some examples:

Satellite CCRMA (http://stanford.io/1riPJsE)
This distribution from Stanford's Center for Computer Research in Music and Acoustics (CCRMA) is geared toward embedded musical instruments and art installations, as well as effects pedals. It can run other software, but the core is Pure Data Extended, Faust, JackTrip, and ChucK. The original rationale is described in Edgar Berdahl and Wendy Ju's paper "Satellite CCRMA: A Musical Interaction and Sound Synthesis Platform" (*http://bit.ly/1qoI7Wo*).

Volumio (http://volumio.org)
A music player for audiophiles. This project evolved from RaspyFi (*http://www.hifiberry.com/hbdigi*).

Also on the music front, you may want to check out the SunVox modular music platform for the Pi (*http://www.warmplace.ru/ soft/sunvox*).

Retrocomputing and Retrogaming

The Pi was inspired by the inexpensive personal computers of the 1980s, so it seems fitting that there are a number of distributions aimed at nostalgic retrocomputing or gaming:

RISC OS (https://www.riscosopen.org/content)
Boots straight into BASIC!

Retropie (http://blog.petrockblock.com/retropie)
An SD card image and GPIO hardware board that makes it easier to build retrogaming consoles.

PiPlay (http://piplay.org)
A prebuilt distribution for gaming and emulation based on MAME (formerly PiMAME).

It's not an OS, but if you're into retro text adventures, try Frotz:

```
sudo apt-get install frotz
```

Internet of Things

The *Internet of Things* or *IoT* describes the realm of devices connected to the Internet. These can be thermostats, bodyweight scales, and doorbells that can be accessed remotely via the Web or in a mobile app. The realm of IoT extends beyond the home as well. Large companies put their equipment online to monitor their assets, which may be spread all over the globe.

Because of Raspberry Pi's low cost and connectivity, it makes a great choice for experimentation with the Internet of Things. While you could use Raspbian for this (in fact, see "Connecting the Web to the Real World" on page 163), here are a few operating systems that are geared toward IoT.

Snappy Ubuntu Core (https://developer.ubuntu.com/en/snappy)
Available for Raspberry Pi 2, Snappy Ubuntu Core is a Linux distribution aimed at devices and cloud servers. It's a barebones operating system that includes methods for being very specific about which versions of applications are installed to support your project. It's meant to be lean, faster, more reliable, and more secure. While it's very new and not fully supported yet, it's worth keeping an eye on developments with Snappy Ubuntu Core.

Windows 10 IoT Core (https://dev.windows.com/en-us/iot)
With the release of Windows 10 in July 2015, Microsoft doubled down on IoT with the release of the Windows 10 IoT Core. While it doesn't have the full Windows desktop environment, developers can now deploy universal Windows applications to Raspberry Pi 2. If you already have experience developing Windows applications, trying out Windows 10 IoT Core on Raspberry Pi 2 is a no-brainer. If you're interested in getting started, Microsoft provides excellent getting-started documentation.

Other Useful Distributions

Here are a few other distributions of note:

Qt on Pi (http://www.qt.io)
An OS bundle aimed at developers of standalone "single-purpose appliances" using the Qt GUI framework.

Web kiosk (http://www.binaryemotions.com)
For making Internet kiosks and digital signage.

Openwrt (https://openwrt.org)
Turn your Pi into a powerful router with this open source router platform.

OctoPrint (http://octoprint.org)
With OctoPrint, you can control your 3D printer via your network. Just connect it to a Raspberry Pi and boot from the OctoPi SD card image.

PiNet (http://pinet.org.uk)
PiNet is a way for teachers to manage a classroom full of Raspberry Pis. It's a method to boot the Pis from a central server on a network. Student accounts are stored on the server, and there are classroom-specific tools preloaded.

Going Further

List of Linux distributions that work with Raspberry Pi (http://bit.ly/1BW4SGp)
This is the definitive list at the Raspberry Pi Hub.

Raspberry Pi Downloads (https://www.raspberrypi.org/downloads)
The downloads section of the Raspberry Pi Foundation website will have a listing of notable operating systems and distributions.

4/Python on the Pi

Python is a great first programming language; it's clear and easy to get up and running. More important, there are a lot of other users to share code with and ask questions.

Guido van Rossum created Python, and very early on recognized its use as a first language for computing. In 1999, van Rossum put together a widely read proposal called "Computer Programming for Everybody" (*http://www.python.org/doc/essays/cp4e/*) that laid out a vision for an ambitious program to teach programming in elementary and secondary schools using Python. More than a decade later, it looks like it is actually happening with the coming of the Raspberry Pi.

Python is an *interpreted language*, which means that you can write a program or script and execute it directly rather than compiling it into machine code. Interpreted languages are a bit quicker to program with, and you get a few side benefits. For example, in Python you don't have to explicitly tell the computer whether a variable is a number, a list, or a string; the interpreter figures out the data types when you execute the script.

The Python interpreter can be run in two ways: as an interactive shell to execute individual commands, or as a command-line program to execute standalone scripts. The integrated development environment (IDE) bundled with Python and the Raspberry Pi is called IDLE.

The Python Version Conundrum

The reason you see two versions of IDLE is that there are two versions of Python installed on the Pi. This is common practice (though a bit confusing). As of this writing, Python 3 is the newest, but changes made to the language between versions 2 and 3 made the latter not backward compatible. Even though Python 3 has been around for years, it took a while for it to be widely adopted, and lots of user-contributed packages have not been upgraded to Python 3. It gets even more confusing when you search the Python documentation; make sure you're looking at the right version!

The default Python version on the Pi is 2.7. You can explicitly run Python 3 with:

```
python3
```

The examples in this book will work with Python 2.7 or 3.X, unless otherwise noted.

Hello, Python

The best way to start learning Python is to jump right in. Although you can use any text editor to start scripting, we'll start out using the IDE. Open up the IDLE 3 application. To run IDLE, click the desktop menu in the lower left, and choose Programming→Python (IDLE 3).

When IDLE opens you'll see a window with the interactive shell. The triple chevron (>>>) is the interactive prompt; when you see the prompt, it means the interpreter is waiting for your commands. At the prompt, type the following:

```
>>> print("Saluton Mondo!")
```

Hit Enter or Return. Python executes that statement and you'll see the result in the shell window. You can use the shell as a kind of calculator to test out statements or calculations:

```
>>> 3+4+5
12
```

Think of the statements executed in the interactive shell as a program that you're running one line at a time. You can set up variables or import modules:

```
>>> import math
>>> (1 + math.sqrt(5)) / 2
1.618033988749895
```

The `import` command makes all of Python's math functions available to your program (more about modules in "Objects and Modules" on page 57). To set up a variable, use the assignment operator (=):

```
>>> import math
>>> radius = 20
>>> radius * 2 * math.pi
1256.6370614359173
```

If you want to clear all variables and start in a fresh state, select Shell→Restart Shell to start over. You can also use the interactive shell to get information about how to use a particular statement, module, or other Python topics with the `help()` command:

```
help("print")
```

To get a listing of all of the topics available, try:

```
help("topics")
help("keywords")
help("modules")
```

The Python interpreter is good for testing statements or simple operations, but you will often want to run your Python script as you would a standalone application. To start a new Python program, select File→New Window, and IDLE will give you a script editing window (see Figure 4-1).

Try typing a line of code and selecting Run→Run Module. You'll get a warning that "Source Must Be Saved OK To Save?". Save your script in your home directory as *SalutonMondo.py* and you'll see it execute in the shell.

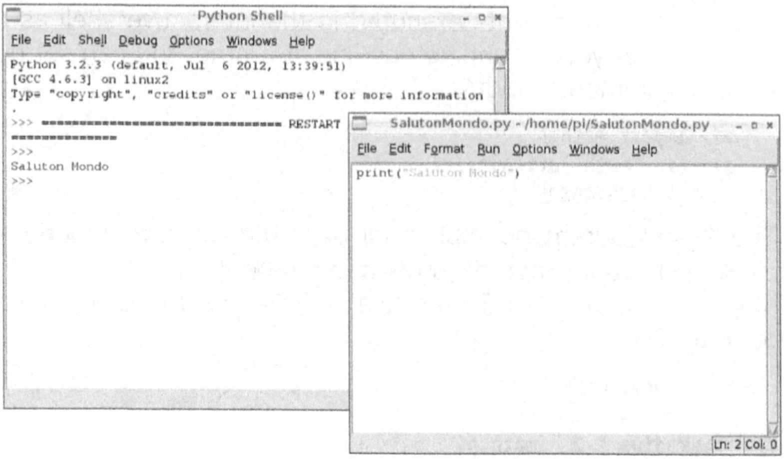

Figure 4-1. *The IDLE interactive shell (left) and an editor window (right)*

Sometimes you may not want to use the IDLE environment. To run a script from the command line, open up the Terminal and type:

```
python SalutonMondo.py
```

That's really all of the basic mechanics you need to know to get up and running with the environment. Next, you'll need to start learning the language.

A Bit More Python

If you're coming to Python from the Arduino world, you're used to writing programs (known as *sketches* in Arduino, but often called *scripts* in Python) in a setup/loop format, where `setup()` is a function run once and `loop()` is a function that executes over and over. The following example shows how to achieve this in Python. Select New Window from the shell in IDLE 3 and type the following:

```
# Setup
n = 0

# Loop
while True:
    n = n + 1
    # The % is the modulo operator
    if ((n % 2) == 0):
        print(n)
```

Select Run Module, and give your script a name (such as *Count-Evens.py*). As it runs, you should see all even integers printed (press Ctrl-C to interrupt the output, because it will go on forever).

You could also implement this using a **for** loop that just counts the first 100 even integers:

```
for n in range(0, 100):
    if ((n % 2) == 0):
        print(n)
```

In the preceding example, you may not notice that each level of indentation is four spaces, not a tab (but you can press Tab in IDLE, and it will dutifully insert spaces for you). Indentation has structural meaning in Python. This is one of the big stumbling blocks for beginners, or when copying and pasting code. Still, we feel that the mandatory use of whitespace makes Python a fairly readable language. See the Python Style Guidelines (*http://bit.ly/1ne4xO5*) for tips on writing readable code.

It is important to watch your whitespace; Python is a highly structured language where the whitespace determines the structure. In the next example, everything indented one level below the **loop()** function is considered part of that function. The end of a loop is determined by where the indentation moves up a level (or end of the file). This differs from languages like C that delimit blocks of code with brackets or other markers.

Use functions to put chunks of code into a code block that can be called from other places in your script. To rewrite the previous example with functions, do the following (when you go to run it, save it as *CountEvens.py*):

```python
# Declare global variables
n = 0   ❶

# Setup function
def setup():   ❷
    global n
    n = 100

def loop():   ❸
    global n
    n = n + 1
    if ((n % 2) == 0):
        print(n)

# Main   ❹
setup()
while True:
    loop()
```

In this example, the output will be every even number from 102 on. Here's how it works:

❶ First, the variable n is defined as a global variable that can be used in any block in the script.

❷ Here, the setup() function is defined (but not yet executed).

❸ Similarly, here's the definition of the loop() function.

❹ In the main code block, setup() is called once, then loop().

The use of the global keyword in the first line of each function is important; it tells the interpreter to use the global variable *n* rather than create a second (local, or private to that function) *n* variable usable only in the function.

This tutorial is too short to be a complete Python reference. To really learn the language, you may want to start with *Learn Python the Hard Way* (*http://learnpythonthehardway.org*), *Think Python*, or the *Python Pocket Reference*. The rest of this chapter will give you enough context to get up and running with the later examples and will map out the basic features and modules available.

Command Line Versus IDLE

One thing you will notice is that the output of IDLE is very slow when running example code that prints to the shell. To get an idea of just how slow it is, keep IDLE open and open a new terminal alongside. In IDLE, run the CountEvens script using Run Module; it will need the head start. Then type the following at the terminal's command-line prompt:

```
python CountEvens.py
```

You'll quickly get an idea of the overhead from using the IDE on the fairly limited resources of the Pi. The examples later in the book will all be executed from the command line, but IDLE can still be used as an editor if you like.

Objects and Modules

You'll need to understand the basic syntax of dealing with objects and modules to get through the examples in this book. Python is a clean language, with just 34 reserved *keywords* (see Table 4-1). These keywords are the core part of the language that let you structure and control the flow of activity in your script. Pretty much everything that isn't a keyword can be considered an *object*. An object is a combination of data and behaviors that has a name. You can change an object's data, retrieve information from it, and even manipulate other objects.

Table 4-1. *Python has just 34 reserved keywords*

Conditionals	Loops	Built-in functions	Classes, modules, functions	Error handling
if	for	print	class	try
else	in	pass	def	def
elif	while	del	global	finally
not	break		break	raise
or	as		nonlocal	assert
and	continue		yield	with

Conditionals	Loops	Built-in functions	Classes, modules, functions	Error handling
is			import	
True			return	
False			from	
None				

In Python, strings, lists, functions, modules, and even numbers are objects. A Python object can be thought of as an encapsulated collection of attributes and methods. You get access to these attributes and methods using a simple dot syntax. For example, type this at the interactive shell prompt to set up a string object and call the method that tells it to capitalize itself:

```
>>> myString = "quux"
>>> myString.capitalize()
'Quux'
```

Or use `reverse()` to rearrange a list in reverse order:

```
>>> myList = ['a', 'man', 'a', 'plan', 'a', 'canal']
>>> myList.reverse()
>>> print(myList)
['canal', 'a', 'plan', 'a', 'man', 'a']
```

Both string and list are built-in modules of the *standard library*, which are available from any Python program. In each case, the string and list modules have defined a bunch of functions for dealing with strings and lists, including `capitalize()` and `reverse()`.

Some of the standard library modules are not built-in, and you need to explicitly say you're going to use them with the `import` command. To use the `time` module from the standard library to gain access to helpful functions for dealing with timing and timestamps, use:

```
import time
```

You may also see `import as` used to rename the module in your program:

```
import time as myTime
```

Or `from import` used to load select functions from a module:

```
from time import clock
```

Here's a short example of a Python script using the `time` and `datetime` modules from the standard library to print the current time once every second:

```
from datetime import datetime
from time import sleep

while True:
    now = str(datetime.now())
    print(now)
    sleep(1)
```

The `sleep` function stops the execution of the program for one second. One thing you will notice after running this code is that the time will drift a bit each time.

That's for two reasons:

- The code doesn't take into account the amount of time it takes to calculate the current time.

- Other processes are sharing the CPU and may take cycles away from your program's execution. This is an important thing to remember: when programming on the Raspberry Pi, you are not executing in a *real-time environment*.

If you're using the `sleep()` function, you'll find that it is accurate to more than 5ms on the Pi.

Next, let's modify the example to open a text file and periodically log some data to it. Everything is a string when handling text files. Use the `str()` function to convert numbers to strings (and `int()` to change back to an integer):

```
from datetime import datetime
from time import sleep
import random

log = open("log.txt", "w")
```

```
for i in range(5):
    now = str(datetime.now())
    # Generate some random data in the range 0-1024
    data = random.randint(0, 1024)
    log.write(now + " " + str(data) + "\n")
    print(".")
    sleep(.9)
log.flush()
log.close()
```

In a real data-logging application, you'll want to make sure you've got the correct date and time set up on your Raspberry Pi, as described in "Setting the Date and Time" on page 41.

Here's another example (*ReadFile.py*) that reads in a filename as an argument from the command line (run it from the shell with **python3 ReadFile.py** *filename*. The program opens the file, reads each line as a string, and prints it. Note that **print()** acts like **println()** does in other languages; it adds a newline to the string that is printed. The **end** argument to **print()** suppresses the newline:

```
# Open and read a file from command-line argument
import sys

if (len(sys.argv) != 2):
    print("Usage: python ReadFile.py filename")
    sys.exit()

scriptname = sys.argv[0]
filename = sys.argv[1]

file = open(filename, "r")
lines = file.readlines()
file.close()

for line in lines:
    print(line, end = '')
```

Even More Modules

One of the reasons Python is so popular is that there are a great number of user-contributed modules that build on the standard library. The Python Package Index (PyPI) (*http://pypi.python.org/pypi*) is the definitive list of packages (or modules); some of the more popular modules that are particularly useful on the Raspberry Pi are shown in Table 4-2. You'll be using some of these modules later on, especially the GPIO module to access the general inputs and outputs of the Raspberry Pi.

Table 4-2. *Some packages of particular interest to Pi users*

Module	Description	URL	Package name
RPi.GPIO	Access to GPIO pins	sourceforge.net/ projects/raspberry-gpio-python	python-rpi.gpio
GPIOzero	Simplified access to GPIO pins	gpiozero.readthedocs.org (*https://gpiozero.readthe docs.org*)	python-gpio-zero
Pygame	Gaming framework	pygame.org	python-pygame
SimpleCV	Easy API for Computer Vision	simplecv.org	No package
SciPy	Scientific computing	www.scipy.org	python-scipy
NumPy	The numerical underpinnings of Scipy	numpy.scipy.org	python-numpy
Flask	Microframework for web development	flask.pocoo.org	python-flask
Feedparser	Atom and RSS feed parser	pypi.python.org/pypi/ feedparser	No package
Requests	"HTTP for Humans"	docs.python-requests.org	python-requests

Module	Description	URL	Package name
PIL	Image processing	*www.pythonware.com/ products/pil/*	python-imaging
wxPython	GUI framework	*wxpython.org*	python-wxgtk2.8
pySerial	Access to serial port	*https://github.com/ pyserial/pyserial*	python-serial
PyUSB	FTDI-USB interface	*bleyer.org/pyusb*	No package

To use one of these, you'll need to download the code, configure the package, and install it. The NumPy module, for example, can be installed as follows:

```
sudo apt-get install python-numpy
```

If a package has been bundled by its creator using the standard approach to bundling modules (with Python's distutils tool), all you need to do is download the package, uncompress it, and type:

```
python setup.py install
```

Easy Module Installs with Pip

Many modules can be installed with apt-get. You may also want to look at the Pip package installer (*https://pip.pypa.io*), a tool that makes it quite easy to install packages from the PyPI. Install Pip using apt-get:

```
sudo apt-get install python-pip
```

Then you can install most modules using Pip to manage the downloads and dependencies. For example:

```
pip install flask
```

In later chapters, you'll use application-specific modules extensively, but here's one example that shows how powerful some modules can be. The Feedparser module is a universal parser that lets you grab RSS or Atom feeds and easily access the content. Because most streams of information on the Web have

RSS or Atom output, Feedparser is one of several ways to get your Raspberry Pi hooked into the Internet of Things.

First, install the Feedparser module using Pip (see "Easy Module Installs with Pip" on page 62):

```
sudo pip install feedparser
```

To use it, simply give the parse function the URL of an RSS feed. Feedparser will fetch the XML of the feed and parse it, and turn it into a special list data structure called a *dictionary*. A Python dictionary is a list of key/value pairs, sometimes called a *hash* or *associative array*. The parsed feed is a dictionary, and the parsed items in the feed are also a dictionary, as shown in this example, which grabs the current weather in Providence, RI, from *http://weather.gov*:

```
import feedparser

feed_url = "http://w1.weather.gov/xml/current_obs/KPVD.rss"
feed = feedparser.parse(feed_url)
RSSitems = feed["items"]
for item in RSSitems:
    weather = item["title"]
    print weather
```

Launching Other Programs from Python

Python makes it fairly easy to trigger other programs on your Pi with the sys.subprocess module. Try the following:

```
from datetime import datetime
from time import sleep
import subprocess
for count in range(0, 60):
    filename = str(datetime.now()) + ".jpg"
    subprocess.call(["fswebcam", filename])
    sleep(60)
```

This is a simple time-lapse script that will snap a photo from a webcam once a minute for an hour. Cameras are covered in greater detail in Chapter 9, but this program should work with a USB camera connected and the fswebcam program installed; run this command first to install it:

```
sudo apt-get install fswebcam
```

The `subprocess.call` function takes a list of strings, concatenates them together (with spaces between), and attempts to execute the program specified. If it is a valid program, the Raspberry Pi will spawn a separate process for it, which will run alongside the Python script. The command in the preceding example is the same as if you typed the following at a terminal prompt:

```
fswebcam 20160812.jpg
```

To add more parameters to the program invocation, just add strings to the list in the `subprocess.call` function:

```
subprocess.call(["fswebcam", "-r", "1280x720", filename])
```

This will send the `fswebcam` program a parameter to change the resolution of the snapped image.

Running a Python Script Automatically at Startup

Because the Pi can act as a standalone appliance, a common question is how to launch a Python script automatically when the Pi boots up. The answer is to add an entry to the */etc/rc.local* file, which is used for exactly this in the Linux world. Just edit the file:

```
sudo nano /etc/rc.local
```

and add a command to execute your script between the commented section and `exit 0`. Something like:

```
python /home/pi/foo.py&
```

The ampersand at the end will run the script as a background process, which will allow all the other services of the Pi to continue booting up.

Troubleshooting Errors

Inevitably, you'll run into trouble with your code, and you'll need to track down and squash a bug. The IDLE interactive mode can be your friend; the Debug menu provides several tools that will help you understand how your code is actually executing. You also have the option of seeing all your variables and stepping through the execution line by line.

Syntax errors are the easiest to deal with; usually this is just a typo or a misunderstood aspect of the language. *Semantic errors*—where the program is well formed but doesn't perform as expected—can be harder to figure out. That's where the debugger can really help unwind a tricky bug. Effective debugging takes years to learn, but here is a quick cheat sheet of things to check when programming the Pi in Python:

- Use `print()` to show when the program gets to a particular point.
- Use `print()` to show the values of variables as the program executes.
- Double-check whitespace to make sure blocks are defined the way you think they are.
- When debugging syntax errors, remember that the actual error may have been introduced well before the interpreter reports it.
- Double-check all of your global and local variables.
- Check for matching parentheses.
- Make sure the order of operations is correct in calculations; insert parentheses if you're not sure. For example, `3 + 4 * 2` and `(3 + 4) * 2` yield different results.

After you're comfortable and experienced with Python, you may want to look at the `code` and `logging` modules for more debugging tools.

Going Further

There is a lot more to Python, and here are some resources that you'll find useful:

Think Python, *by Allen Downey*
> This is a clear and fairly concise approach to programming (that happens to use Python).

Python Pocket Reference, *by Mark Lutz*
> Because sometimes flipping through a book is better than clicking through a dozen Stack Overflow posts.

Stack Overflow (http://stackoverflow.com)
> That said, Stack Overflow is an excellent source of collective knowledge. It works particularly well if you're searching for a specific solution or error message; chances are someone else has had the same problem and posted here.

Learn Python the Hard Way (http://learnpythonthehardway.org), *by Zed Shaw*
> A great book and online resource; at the very least, read the introduction "The Hard Way Is Easier."

Python for Kids, *by Jason R. Briggs*
> Again, more of a general programming book that happens to use Python (and written for younger readers).

5/Arduino and the Pi

As you'll see in the next few chapters, you can use the GPIO pins on the Raspberry Pi to connect to sensors or things like blinking LEDs and motors. And if you have experience using the Arduino microcontroller development platform, you can also use that alongside the Raspberry Pi.

When the Raspberry Pi was first announced, a lot of people asked if it was an Arduino killer. For about the same price, the Pi provides much more processing power: why use Arduino when you have a Pi? It turns out the two platforms are actually complementary, and the Raspberry Pi makes a great host for the Arduino. There are quite a few situations where you might want to put the Arduino and Pi together:

- To use the large number of libraries and sharable examples for the Arduino.

- To supplement an Arduino project with more processing power. For example, maybe you have a MIDI controller that was hooked up to a synthesizer, but now you want to upgrade to synthesizing the sound directly on the Pi.

- When you're dealing with 5V logic levels. The Pi operates at 3.3V, and its pins are not tolerant of 5V. The Arduino can act as a "translator" between the two.

- To prototype something a little out of your comfort zone, in which you may make some chip-damaging mistakes. For example, we've seen students try to drive motors directly from a pin on the Arduino (don't try it); it was easy to pry the damaged microcontroller chip out of its socket and replace it (less than $10 usually). Not so with the Raspberry Pi.

- When you have a problem that requires exact control in real time, such as a controller for a 3D printer. As we saw in Chapter 4, Raspbian is not a real-time operating system, and programs can't necessarily depend on the same "instruction per clock cycles" rigor of a microcontroller.

The examples in this section assume that you know at least the basics of using the Arduino development board and integrated development environment (IDE). If you don't have a good grasp of the fundamentals, *Getting Started with Arduino* by Massimo Banzi is a great place to start. The official Arduino tutorials (*http://bit.ly/1oTWBNB*) are quite good as well, and provide a lot of opportunities to cut and paste good working code.

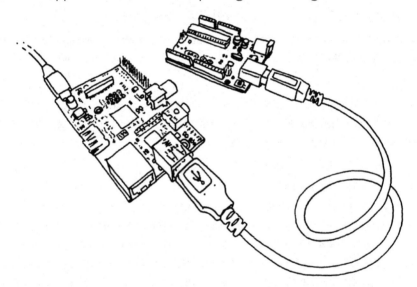

Figure 5-1. *Arduino and the Raspberry Pi are BFFs*

Installing Arduino in Raspbian

To program an Arduino development board, you need to hook it up to a computer with a USB cable, then compile and flash a program to the board using the Arduino IDE. You can do this with any computer, or you can use your Raspberry Pi as a host to program the Arduino.

Using the Raspberry Pi to program the Arduino will be quicker to debug, but compiling will be a little slower on the Pi than on a modern laptop or desktop computer. It's not too bad though, and you'll find that compiling will take less time after the very first compile, as Arduino only compiles code that has changed since the last compilation.

To install the Arduino IDE on the Raspberry Pi, type the following into a terminal:

```
sudo apt-get update  ❶
sudo apt-get install arduino  ❷
```

❶ Make sure you have the latest package list.

❷ Download the Arduino package.

This command will install Java plus a lot of other dependencies. The Arduino environment will appear under the *Electronics* section of the program menu (don't launch it just yet though).

You can just plug the Arduino into one of the open USB ports. The USB connection will be able to provide enough power for the Arduino, but you might want to power the Arduino separately, depending on your application (if you're running motors or heaters for instance).

 Note that you'll need to plug the Arduino USB cable in after the Raspberry Pi has booted up. If you leave it plugged in at boot time, the Raspberry Pi may hang as it tries to figure out all the devices on the USB bus.

When you launch the Arduino IDE, it polls all the USB devices and builds a list that is shown in the Tools→Serial Port menu. Click Tools→Serial Port and select the serial port (most likely /dev/ttyACM0), then click Tools→Board, and select the type of Arduino Board you have (e.g., Uno). Click File→Examples→01.Basics→Blink to load a basic example sketch. Click the Upload button in the toolbar or choose File→Upload to upload the sketch, and after the sketch loads, the Arduino's onboard LED will start blinking.

In order to access the serial port on versions of Raspbian older than Jessie, you'll need to make sure that the *pi* user has permission to do so. You don't have to do this step on Jessie. You can do that by adding the *pi* user to the *tty* and *dialout* groups. You'll need to do this before running the Arduino IDE:

```
sudo usermod ❶ -a -G ❷ tty pi
sudo usermod -a -G dialout pi
```

❶   usermod is a Linux program to manage users.

❷   -a -G puts the user (pi) in the specified group (tty, then dialout).

Finding the Serial Port

If, for some reason, /dev/ttyACM0 doesn't work, you'll need to do a little detective work. To find the USB serial port that the Arduino is plugged into without looking at the menu, try the following from the command line. Without the Arduino connected, type:

```
ls /dev/tty*
```

Plug in Arduino, then try the same command again and see what changed. On my Raspberry Pi, at first I had /dev/ttyAMA0 listed (which is the onboard USB hub). When I plugged in the Arduino, /dev/ttyACM0 popped up in the listing.

Improving the User Experience

While you're getting set up, you may notice that the quality of the default font in the Arduino editor is less than ideal. You can improve it by downloading the open source font Inconsolata. To install (when the Arduino IDE is closed), type:

```
sudo apt-get install fonts-inconsolata
```

Then edit the Arduino preferences file:

```
nano ~/.arduino/preferences.txt
```

and change the following lines to:

```
editor.font=Inconsolata,medium,14
editor.antialias=true
```

When you restart Arduino, the editor will use the new font.

Talking in Serial

To communicate between the Raspberry Pi and the Arduino over a serial connection, you'll use the built-in *Serial* library on the Arduino side, and the Python pySerial (*https://github.com/pyserial/pyserial*) module on the Raspberry Pi side. It comes pre-installed on Jessie, but if you ever need to install it, the package names are:

```
sudo apt-get install python-serial python3-serial
```

Open the Arduino IDE and upload this code to the Arduino:

```
void setup() {
  Serial.begin(9600);
}

void loop() {
  for (byte n = 0; n < 255; n++) {
    Serial.write(n);
    delay(50);
  }
}
```

This counts upward and sends each number over the serial connection.

Note that in Arduino, `Serial.write()` sends the actual number, which will get translated on the other side as an ASCII character code. If you want to send the string "123" instead of the number 123, use the `Serial.print()` command.

Next, you'll need to know which USB serial port the Arduino is connected to (see "Finding the Serial Port" on page 70). Here's the Python script; if the port isn't */dev/ttyACM0*, change the value of **port**. (See Chapter 4 for more on Python). Save it as *SerialEcho.py* and run it with `python SerialEcho.py`:

```
import serial

port = "/dev/ttyACM0"
serialFromArduino = serial.Serial(port,9600)   ❶
serialFromArduino.flushInput()   ❷
while True:
    if (serialFromArduino.inWaiting() > 0):
        input = serialFromArduino.read(1)   ❸
        print(ord(input))   ❹
```

❶ Open the serial port connected to the Arduino.

❷ Clear out the input buffer.

❸ Read one byte from the serial buffer.

❹ Change the incoming byte into an actual number with `ord()`.

You won't be able to upload to Arduino when Python has the serial port open, so make sure you kill the Python program with Ctrl-C before you upload the sketch again. You will be able to upload to an Arduino Leonardo or Arduino Micro, but doing so will break the connection with the Python script, so you'll need to restart it anyhow.

The Arduino is sending a number to the Python script, which interprets that number as a string. The `input` variable will contain whatever character maps to that number in the ASCII table (*http://bit.ly/ZS47DO*). To get a better idea, try replacing the last line of the Python script with this:

```
print(str(ord(input)) + " = the ASCII character " + input +
".")
```

Setting the Serial Port as an Argument

If you want to set the port as a command-line argument, use the `sys` module to grab the first argument:

```
import serial, sys

if (len(sys.argv) != 2):
    print("Usage: python ReadSerial.py port")
    sys.exit()
port = sys.argv[1]
```

After you do this, you can run the program like this:

```
python SerialEcho.py /dev/ttyACM0
```

The first simple example just sent a single byte; this could be fine if you are only sending a series of event codes from the Arduino. For example, if the left button is pushed, send a 1; if the right, send 2. That's only good for 255 discrete events, though; more often you'll want to send arbitrarily large numbers or strings. If you're reading analog sensors with the Arduino, for example, you'll want to send numbers in the range 0 to 1,023.

Parsing arbitrary numbers that come in one byte at a time can be trickier than you might think in many languages. The way Python and pySerial handle strings makes it almost trivial, however. As a simple example, update your Arduino with the following code that counts from 0 to 1,024:

```
void setup() {
  Serial.begin(9600);
}
```

```
void loop() {
  for (int n = 0; n < 1024; n++)
    Serial.println(n, DEC);
    delay(50);
  }
}
```

The key difference is in the **println()** command. In the previous example, the **Serial.write()** function was used to write the raw number to the serial port. With **println()**, the Arduino formats the number as a decimal string and sends the ASCII codes for the string. So instead of sending a byte with the value **254**, it sends the string **254\r\n**. The **\r** represents a carriage return, and the **\n** represents a newline (these are concepts that carried over from the typewriter into computing: carriage return moves to the start of the line; newline starts a new line of text).

On the Python side, you can use **readline()** instead of **read()**, which will read all of the characters up until (and including) the carriage return and newline. Python has a flexible set of functions for converting between the various data types and strings. It turns out you can just use the **int()** function to change the formatted string into an integer:

```
import serial

port = "/dev/ttyACM0"
serialFromArduino = serial.Serial(port,9600)
serialFromArduino.flushInput()
while True:
    input = serialFromArduino.readline()
    inputAsInteger = int(input)
    print(inputAsInteger * 10)
```

Note that it is simple to adapt this example so that it will read an analog input and send the result; just change the loop to:

```
void setup() {
  Serial.begin(9600);
}

void loop() {
  int n = analogRead(A0);
  Serial.println(n, DEC);
  delay(100);
}
```

Assuming you change the Python script to just print `inputAsInteger` instead of `inputAsInteger * 10`, you should get some floating values in the 200 range if nothing is connected to analog pin 0. With some jumper wire, connect pin 0 to GND and the value should be 0. Connect it to the 3V3 pin and you'll see a value around 715, and 1,023 when connected to the 5V pin.

Using Arduino Compatibles

There are many microcontroller boards that are compatible with the Arduino IDE. Some use a special adapter from FTDI that handles all of the USB-to-TTL serial communication. To connect to these (often more inexpensive) boards, you would use an FTDI cable or an adapter board like the USB BUB or FTDI Friend. In the Jessie release, the FTDI driver for these boards is already installed, so they should work right out of the box. These boards typically show up as the */dev/ttyUSB0* device.

Using Firmata

As you go deeper, many projects will look the same as far as the code for basic communication goes. As with any form of communication, things get tricky once you get past "Hello, world"; you'll need to create *protocols* (or find an existing protocol and implement it) so that each side understands the other.

One good solution that comes bundled with Arduino is Hans-Christoph Steiner's Firmata (*http://bit.ly/1CXFnqF*), an all-purpose serial protocol that is simple and human readable. It may not be perfect for all applications, but it is a good place to start. Here's a quick example:

1. Select File→Examples→Firmata→StandardFirmata in Arduino; this will open the sample Firmata code that will allow you to send and receive messages to the Arduino and get information about all of the pins.

2. Upload that code to the Arduino the same way you did in previous examples.

3. You'll need a bit of Python code on the Pi to send commands to the Arduino to query or change its state. The easiest way

is to use the pyFirmata module. Install it using Pip (see "Easy Module Installs with Pip" on page 62):

```
sudo pip install pyfirmata
```

4. Because Firmata is a serial protocol, you talk to the Arduino from Python in the same way as in previous examples, but using pyFirmata instead of pySerial. Use the **write()** method to make a digital pin high or low on the Arduino:

```
from pyfirmata import Arduino
from time import sleep
board = Arduino('/dev/ttyACM0')
while (1):
    board.digital[13].write(1)
    print("on")
    sleep(1)
    board.digital[13].write(0)
    print("off")
    sleep(1)
```

This code should blink the LED on the Arduino Uno board that is connected to pin 13. The full module is documented on the pyFirmata GitHub page (*https://github.com/tino/pyFirmata*).

Going Further

The nitty-gritty of serial protocols is beyond the scope of this book, but there are a lot of interesting examples of how other people have solved problems in the "Interfacing with Software" (*http://bit.ly/1o17nGY*) section of the Arduino Playground (*http://www.arduino.cc/playground*). In addition, you may want to try:

MIDI
  If your project is musical, consider using MIDI commands as your serial protocol. MIDI is (basically) just serial, so it should just work.

Arduino-compatible Raspberry Pi shields
  There are a few daughterboards (or shields) on the market that connect the GPIO pins on the Raspberry Pi with an Arduino-compatible microcontroller. WyoLum's AlaMode

(*http://bit.ly/1EylgRM*) shield is a good solution and offers a few other accessories, including a real-time clock.

Talk over a network
Finally, you can ditch the serial connection altogether and talk to the Arduino over a network. A lot of really interesting projects are using the WebSocket (*http://www.websocket.org*) protocol along with the Node.js (*http://nodejs.org*) JavaScript platform.

Using the serial pins on the Raspberry Pi header
The header on the Raspberry Pi pulls out a number of input and output pins, including two that can be used to send and receive serial data bypassing the USB port. To do that, you'll first need to cover the material in Chapter 7, and make sure that you have a level shifter to protect the Raspberry Pi 3.3V pins from the Arduino's 5V pins.

If you're looking to get deeper into making physical devices communicate, a good starting point is *Making Things Talk, 2nd Edition* (*http://bit.ly/1oTXntU*), by Tom Igoe.

6/Basic Input and Output

While the Raspberry Pi is, in essence, a very inexpensive Linux computer, there are a few things that distinguish it from laptop and desktop machines that we usually use for writing email, browsing the Web, or word processing. One of the main differences is that the Raspberry Pi can be directly used in electronics projects, because it has *general-purpose input/output* pins right on the board, shown in Figure 6-1.

Figure 6-1. *Raspberry Pi's GPIO pins*

These GPIO pins can be accessed for controlling hardware such as LEDs, motors, and relays, which are all examples of outputs. As for inputs, your Raspberry Pi can read the status of buttons, switches, and dials, or it can read sensors for things like temperature, light, motion, or proximity (among many others).

With the introduction of Raspberry Pi 1 Model B+, the number of GPIO pins increased from 26 to 40. If you have an older Raspberry Pi, you can still carry out the examples in this chapter as they'll use only the first 26 pins on the GPIO header.

The best part of having a computer with GPIO pins is that you can create programs to read the inputs and control the outputs based on many different conditions, as easily as you'd program your desktop computer. Unlike a typical microcontroller board, which also has programmable GPIO pins, the Raspberry Pi has a few extra inputs and outputs, such as your keyboard, mouse, and monitor, as well as the Ethernet port, which can act as both an input and an output. If you have experience creating electronics projects with microcontroller boards like the Arduino, you have a few more inputs and outputs at your disposal with the Raspberry Pi. Best of all, they're built right in; there's no need to wire up any extra circuitry to use them.

Having a keyboard, mouse, and monitor is not the only advantage that Raspberry Pi has over typical microcontroller boards. There are a few other key features that will help you in your electronics projects:

Filesystem
    Being able to read and write data in the Linux filesystem will make many projects much easier. For instance, you can connect a temperature sensor to the Raspberry Pi and have it take a reading once a second. Each reading can be appended to the end of a log file, which can be easily downloaded and parsed in a graphing program. It can even be graphed right on the Raspberry Pi itself!

Linux tools

Packaged in the Raspberry Pi's Linux distribution is a set of core command-line utilities, which let you work with files, control processes, and automate many different tasks. These powerful tools are at your disposal for all of your projects. And because there is an enormous community of Linux users that depend on these core utilities, getting help is usually one web search away. For general Linux help, you can usually find answers at Stack Overflow (*http://stacko verflow.com*). If you have a question specific to Raspberry Pi, try the Raspberry Pi Forum (*http://www.raspberrypi.org/phpBB3*) or the Raspberry Pi section of Stack Overflow (*http://raspberrypi.stackexchange.com*).

Languages

There are many programming languages out there, and embedded Linux systems like the Raspberry Pi give you the flexibility to choose whichever language you're most comfortable with. The examples in this book use shell scripting and Python, but they could easily be translated to languages like C, Java, or Perl.

One of the drawbacks to the Raspberry Pi is that there's no way to directly connect *analog sensors*, such as light and temperature sensors. Doing so requires a chip called an *analog-to-digital converter* or *ADC*. See Chapter 8 for how to read analog sensors using an ADC.

Using Inputs and Outputs

There are a few supplies that you'll need in addition to the Raspberry Pi itself in order to try out these basic input and output tutorials. Many of these parts you'll be able to find in hobby electronics component stores, or they can be ordered online from stores like MakerShed, SparkFun, Adafruit, Mouser, or Digi-Key. Here are a few of the basic parts:

- Solderless breadboard
- LEDs, assorted
- Male-to-male jumper wires
- Female-to-male jumper wires (these are not as common as their male-to-male counterparts but are needed to connect the Raspberry Pi's GPIO pins to the breadboard)
- Pushbutton switch
- Resistors, assorted

To make it easier to connect breadboarded components to the Raspberry Pi's pins, we also recommend Adafruit's Pi Cobbler Breakout Kit, which connects all the GPIO pins to a breadboard with a single ribbon cable. This eliminates the need to use female-to-male jumper wires. If you go with the less-expensive, unassembled version of the kit, it's up to you to solder the parts onto the board, but it's easy to do, and Adafruit's tutorial (*http://bit.ly/1sfqg8u*) walks you through the process step by step.

In Figure 6-2, we've labeled each pin according to its default GPIO signal number, which is how you'll refer to a particular pin in the commands you execute and in the code that you write. The unlabeled pins are assigned to other functions by default.

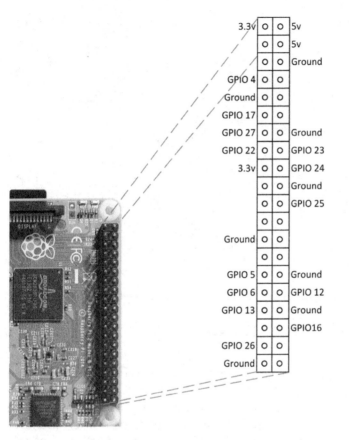

3.3v	o o	5v
	o o	5v
	o o	Ground
GPIO 4	o o	
Ground	o o	
GPIO 17	o o	
GPIO 27	o o	Ground
GPIO 22	o o	GPIO 23
3.3v	o o	GPIO 24
	o o	Ground
	o o	GPIO 25
	o o	
Ground	o o	
	o o	
GPIO 5	o o	Ground
GPIO 6	o o	GPIO 12
GPIO 13	o o	Ground
	o o	GPIO16
GPIO 26	o o	
Ground	o o	

Figure 6-2. *The default GPIO pins on the Raspberry Pi. Some of the pins left blank could also be used as GPIO, but they have other possible functions. Unless you need more GPIO pins than are listed here, steer clear of them for now.*

There's a handy website created by Phil Howard called Raspberry Pinout (*http://pinout.xyz*) which we recommend you bookmark. It'll show you the Raspberry Pi's GPIO pins and has tons of reference information about how they can be used.

There are also great products such as RasPiO Portsplus Port ID board (*http://rasp.io*), shown in Figure 6-3. It's a small board that fits over the GPIO pins for the sole purpose of making the pins easy to identify.

Figure 6-3. *The RasPiO Portsplus Port ID board*

Digital Output: Lighting Up an LED

The easiest way to use outputs with the GPIO pins is by connecting an LED, or light-emitting diode. You can then use the
Linux command line to turn the LED on and off. Once you have
an understanding of how these commands work, you're one
step closer to having an LED light up to indicate when you have
new email, when you need to take an umbrella with you as you
leave your house, or when it's time to go to bed. It's also very
easy to go beyond a basic LED and use a relay to control a lamp
on a set schedule, for instance.

Beginner's Guide to Breadboarding

If you've never used a breadboard (Figure 6-4) before, it's
important to know which terminals are connected. In the diagram, we've shaded the terminal connections on a typical
breadboard. Note that the power buses on the left side are not
connected to the power buses on the right side of the breadboard. You'll have to use male-to-male jumper cables to connect
them to each other if you need ground and voltage on both sides
of the breadboard.

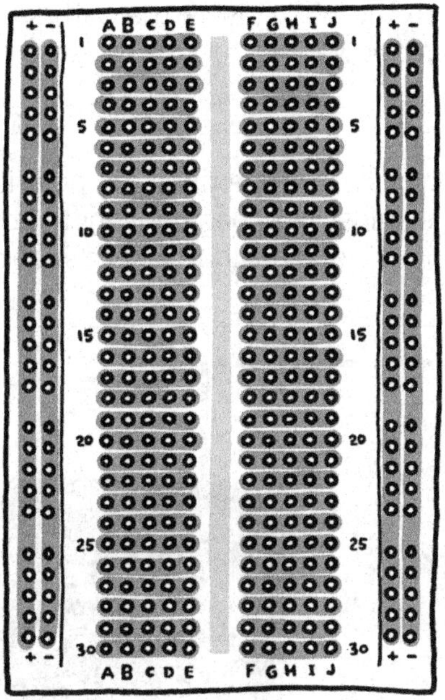

Figure 6-4. *Breadboard*

Here are the instructions you should follow:

1. Using a male-to-female jumper wire, connect pin 25 on the Raspberry Pi to the breadboard. Refer to Figure 6-2 for the location of each pin on the Raspberry Pi's GPIO header.

2. Using another jumper wire, connect the Raspberry Pi's ground pin to the negative power bus on the breadboard.

3. Now you're ready to connect the LED (see Figure 6-5). Before you do that, it's important to know that LEDs are *polarized*: it matters which of the LED's wires is connected to what. Of the two leads coming off the LED, the longer one is the anode (or "plus") and should be connected to a GPIO pin. The shorter lead is the cathode (or "minus") and should be connected to ground. Another way to tell the difference is by looking from the top. The flat side of the LED indicates the cathode, the side that should be connected to ground. Insert the anode side of the LED into the breadboard in the same channel as the jumper wire from pin 25, which will con-

nect pin 25 to the LED. Insert the cathode side of the LED into the ground power bus.

An easy way to remember the polarity of an LED is that the "plus" lead has had length *added* to it, while the "minus" lead has had length *subtracted* from it.

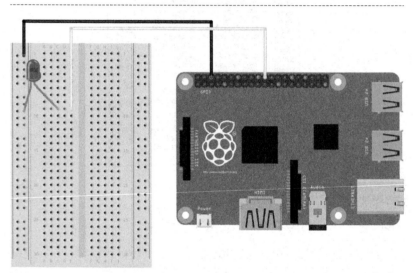

fritzing

Figure 6-5. *Connecting an LED to the Raspberry Pi*

4. With your keyboard, mouse, and monitor hooked up, power on your Raspberry Pi and log in. If you're at a command line, you're ready to go. If you're in the X Window environment, double-click the LXTerminal icon on your desktop. This will bring up a terminal window.

5. In order to access the input and output pins from the command line, you'll need to run the commands as root, the *superuser* account on the Raspberry Pi. To start running commands as root, type **sudo su** at the command line and press Enter:

```
pi@raspberrypi ~ $ sudo su
root@raspberrypi:/home/pi#
```

You'll notice that the command prompt has changed from $ to #, indicating that you're now running commands as root.

The root account has administrative access to all the functions and files on the system and there is very little protecting you from damaging the operating system if you type a command that can harm it. You must exercise caution when running commands as root. If you do mess something up, don't worry about it too much; you can always reimage the SD card with a clean Linux install; however, you'll lose any customization you made to the operating system, as well as any programs or sketches you wrote.

When you're done working within the root account, type **exit** to return to working within the **pi** user account.

6. Before you can use the command line to turn the LED on pin 25 on and off, you need to *export the pin to the userspace* (in other words, make the pin available for use outside of the confines of the Linux kernel), this way:

```
root@raspberrypi:/home/pi# echo 25 > /sys/class/gpio/export
```

The echo command writes the number of the pin you want to use (25) to the export file, which is located in the folder */sys/class/gpio*. When you write pin numbers to this special file, it creates a new directory in */sys/class/gpio* that has the control files for the pin. In this case, it created a new directory called */sys/class/gpio/gpio25*.

7. Change to that directory with the cd command and list the contents of it with ls:

```
root@raspberrypi:/home/pi# cd /sys/class/gpio/gpio25
root@raspberrypi:/sys/class/gpio/gpio25# ls
active_low  direction  edge  power  subsystem  uevent
  value
```

The command `cd` stands for "change directory." It changes the working directory so that you don't have to type the full path for every file. `ls` will list the files and folders within that directory. There are two files that you're going to work with in this directory: `direction` and `value`.

8. The `direction` file is how you'll set this pin to be an input (like a button) or an output (like an LED). Because you have an LED connected to pin 25 and you want to control it, you're going to set this pin as an output:

    ```
    root@raspberrypi:/sys/class/gpio/gpio25# echo out >
    direction
    ```

9. To turn the LED on, you'll use the `echo` command again to write the number 1 to the `value` file:

    ```
    root@raspberrypi:/sys/class/gpio/gpio25# echo 1 > value
    ```

10. After pressing Enter, the LED will turn on! Turning it off is as simple as using `echo` to write a zero to the value file:

    ```
    root@raspberrypi:/sys/class/gpio/gpio25# echo 0 > value
    ```

Virtual Files

The files that you're working with aren't actually files on the Raspberry Pi's SD card, but rather are a part of Linux's *virtual filesystem*, which is a system that makes it easier to access low-level functions of the board in a simpler way. For example, you could turn the LED on and off by writing to a particular section of the Raspberry Pi's memory, but doing so would require more coding and more caution.

So if writing to a file is how you control components that are outputs, how do you check the status of components that are inputs? If you guessed "reading a file," then you're absolutely right. Let's try that now.

Digital Input: Reading a Button

Simple pushbutton switches like the one in Figure 6-6 are great for controlling basic digital input. Best of all, they're made to fit perfectly into a breadboard.

These small buttons are very commonly used in electronics projects and understanding what's going on inside of them will help you as you prototype your project. When looking at the button as it sits in the breadboard (see Figure 6-6), the top two terminals are always connected to each other. The same is true for the bottom two terminals; they're always connected. When you push down on the button, these two sets of terminals are connected to each other.

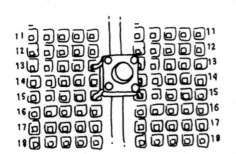

Figure 6-6. *Button*

When you read a digital input on a Raspberry Pi, you're checking to see if the pin is connected to either 3.3 volts or to ground. It's important to remember that it must be either one or the other, and if you try to read a pin that's not connected to either 3.3 volts or ground, you'll get unexpected results. Once you understand how digital input with a pushbutton works, you can start using components like magnetic security switches, arcade joysticks, or even vending machine coin slots. Start by wiring up a switch to read its state:

1. Insert the pushbutton into the breadboard so that its leads straddle the middle channel.

2. Using a jumper wire, connect pin 24 from the Raspberry Pi to one of the top terminals of the button.

3. Connect the 3V3 pin from the Raspberry Pi to the positive power bus on the breadboard.

> Be sure that you connect the button to the 3V3 pin and not the 5V pin. Using more than 3.3 volts on an input pin will permanently damage your Raspberry Pi.

---

4. Connect one of the bottom terminals of the button to the power bus. Now when you push down on the button, the 3.3 volts will be connected to pin 24.

5. Remember what we said about how a digital input must be connected to *either* 3.3 volts or ground? When you let go of the button, pin 24 isn't connected to either of those and is therefore *floating*. This condition will cause unexpected results, so let's fix that. Use a 10K resistor (labeled with the colored bands: brown, black, orange, and then silver or gold) to connect the input side of the switch to the ground rail, which you connected to the Raspberry Pi's ground in the output example. When the switch is not pressed, the pin will be connected to ground.

   Electricity always follows the path of least resistance toward ground, so when you press the switch, the 3.3 volts will go toward the Raspberry Pi's input pin, which has less resistance than the 10K resistor. When everything's hooked up, it should look like Figure 6-7.

6. Now that the circuit is built, let's read the value of the pin from the command line. If you're not already running commands as root, type **sudo su**.

7. As with the previous example, you need to export the input pin to userspace:

```
# echo 24 > /sys/class/gpio/export
```

8. Let's change to the directory that was created during the export operation:

```
# cd /sys/class/gpio/gpio24
```

9. Now set the direction of the pin to input:

```
# echo in > direction
```

fritzing

Figure 6-7. *Connecting a button to the Raspberry Pi*

10. To read the value of the of the pin, you'll use the `cat` command, which will print the contents of files to the terminal. The command `cat` gets its name because it can also be used to concatenate, or join, files. It can also display the contents of a file for you:

```
# cat value
0
```

11. The zero indicates that the pin is connected to ground. Now press and hold the button while you execute the command again:

```
# cat value
1
```

12. If you see the number 1, you'll know you've got it right!

> To easily execute a command that you've previously executed, hit the up arrow key until you see the command that you want to run and then hit Enter.

Now that you can use the Linux command line to control an LED or read the status of a button, let's use a few of Linux's built-in tools to create a very simple project that uses digital input and output.

Project: Cron Lamp Timer

Let's say you're leaving for a long vacation early tomorrow morning and you want to ward off would-be burglars from your home. A lamp timer is a good deterrent, but hardware stores are closed for the night and you won't have time to get one before your flight in the morning. However, because you're a Raspberry Pi hobbyist, you have a few supplies lying around, namely:

- Raspberry Pi board
- Breadboard
- Jumper wires, female-to-male
- PowerSwitch Tail II relay
- Hookup wire

With these supplies, you can make your own programmable lamp timer using two powerful Linux tools: *shell scripts* and *cron*.

Scripting Commands

A shell script is a file that contains a series of commands (just like the ones you've been using to control and read the pins). Take a look at the following shell script and the explanation of the key lines:

```
#!/bin/bash    ❶
echo Exporting pin $1.    ❷
echo $1 > /sys/class/gpio/export    ❸
```

```
echo Setting direction to out.
echo out > /sys/class/gpio/gpio$1/direction  ❹
echo Setting pin high.
echo 1 > /sys/class/gpio/gpio$1/value
```

❶ This line is required for all shell scripts.

❷ $1 refers to the first command-line argument.

❸ Instead of exporting a specific pin number, the script uses the first command-line argument.

❹ Notice that the first command-line argument replaces the pin number here as well.

Save that as a text file called *on.sh* and make it executable with the chmod command:

```
root@raspberrypi:/home/pi# chmod +x on.sh
```

You still need to be executing these commands as root. Type sudo su if you're getting errors like "Permission denied."

A command-line argument is a way of passing information into a program or script by typing it in after name of the command. When you're writing a shell script, $1 refers to the first command-line argument, $2 refers to the second, and so on. In the case of on.sh, you'll type in the pin number that you want to export and turn on. Instead of *hard coding* pin 25 into the shell script, it's more universal by referring to the pin that was typed in at the command line. To export pin 25 and turn it on, you can now type:

```
root@raspberrypi:/home/pi/# ./on.sh 25  ❶
Exporting pin 25.
Setting direction to out.
Setting pin high.
```

❶ The ./ before the filename indicates that you're executing the script in the directory you're in.

If you still have the LED connected to pin 25 from earlier in the chapter, it should turn on. Let's make another shell script called *off.sh*, which will turn the LED off. It will look like this:

```
#!/bin/bash
echo Setting pin low.
echo 0 > /sys/class/gpio/gpio$1/value
echo Unexporting pin $1
echo $1 > /sys/class/gpio/unexport
```

Now let's make it executable and run the script:

```
root@raspberrypi:/home/pi/temp# chmod +x off.sh
root@raspberrypi:/home/pi/temp# ./off.sh 25
Setting pin low.
Unexporting pin 25
```

If everything worked, the LED should have turned off.

Connecting a Lamp

Of course, a tiny little LED isn't going to give off enough light to fool burglars into thinking that you're home, so let's hook up a lamp to the Raspberry Pi:

1. Remove the LED connected to pin 25.

2. Connect two strands of hookup wire to the breadboard, one that connects to pin 25 of the Raspberry Pi and the other to the ground bus.

3. The strand of wire that connects to pin 25 should be connected to the "+in" terminal of the PowerSwitch Tail.

4. The strand of wire that connects to ground should be connected to the "-in" terminal of the PowerSwitch Tail. Compare your circuit to Figure 6-8.

5. Plug the PowerSwitch Tail into the wall and plug a lamp into the PowerSwitch Tail. Be sure the lamp's switch is in the on position.

6. Now when you execute ./on.sh 25, the lamp should turn on; and if you execute ./off.sh 25, the lamp should turn off!

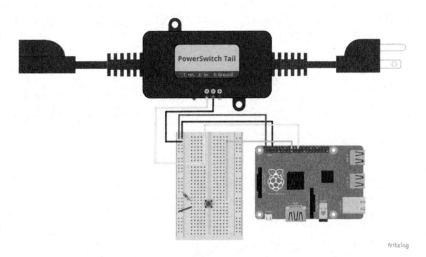

fritzing

Figure 6-8. *Connecting a PowerSwitch Tail II to the Raspberry Pi*

Inside the PowerSwitch Tail there are a few electronic components that help you control high-voltage devices like a lamp or blender by using a low-voltage signal such as the one from the Raspberry Pi. The "click" you hear from the PowerSwitch Tail when it's turned on or off is the relay, the core component of the circuit inside. A relay acts like a switch for the high-voltage device that can be turned on or off depending on whether the low-voltage control signal from the Raspberry Pi is on or off.

Scheduling Commands with cron

So now you've packaged up a few different commands into two simple commands that can turn a pin on or off. And with the lamp connected to the Raspberry Pi through the PowerSwitch Tail, you can turn the lamp on or off with a single command. Now you can use cron to schedule the light to turn on and off at different times of day. cron is Linux's job scheduler. With it, you can set commands to execute on specific times and dates, or you can have jobs run on a particular period (for example, once an hour). You're going to schedule two jobs: one of them will turn the light on at 8:00 p.m., and the other will turn the light off at 2:00 a.m.

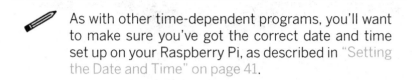

As with other time-dependent programs, you'll want to make sure you've got the correct date and time set up on your Raspberry Pi, as described in "Setting the Date and Time" on page 41.

To add these jobs, you'll have to edit the cron table (a list of commands that Linux executes at specified times):

```
root@raspberrypi:/home/pi/# crontab -e
```

This will launch a text editor to change root's cron table (to change to the root user, type sudo su). At the top of the file, you'll see some information about how to modify the cron table. Use your arrow keys to get to the bottom of the file and add these two entries at the end of the file:

```
0 20 * * * /home/pi/on.sh 25
0 2 * * * /home/pi/off.sh 25
```

cron will ignore any lines that start with the hash mark. If you want to temporarily disable a line without deleting it or add a comment to the file, put a hash mark in front of the line.

Press Ctrl-X to exit, press y to save the file when it prompts you, and hit Enter to accept the default filename. When the file is saved and you're back at the command line, it should say instal ling new crontab to indicate that the changes you've made are going to be executed by cron.

More About cron

cron will let you schedule jobs for specific dates and times or at intervals. There are five time fields (or six if you want to schedule by year), each separated by a space followed by another space, then the command to execute; asterisks indicate that the job should execute each period (Table 6-1).

Table 6-1. *cron entry for turning light on at 8:00 p.m. every day*

0	20	*	*	*	/home/pi/on.sh 25
Minute (:00)	Hour (8 p.m.)	Every day	Every month	Every day of the week	Path to command

Let's say you only wanted the lamp to turn on every weekday. Table 6-2 shows what the `crontab` entry would look like.

Table 6-2. *cron entry for turning light on at 8:00 p.m. every weekday*

0	20	*	*	1-5	/home/pi/on.sh 25
Minute (:00)	Hour (8 p.m.)	Every day	Every month	Monday to Friday	Path to command

As another example, let's say you have a shell script that checks if you have new mail and emails you if you do. Table 6-3 shows how you'd get that script to run every five minutes.

Table 6-3. *cron entry for checking for mail every five minutes*

*/5	*	*	*	*	/home/pi/ checkMail.sh
Every five minutes	Every hour	Every day	Every month	Every day of the week	Path to command

The */5 indicates a period of every five minutes.

As you can see, `cron` is a powerful tool that's at your disposal for scheduling jobs for specific dates or times and at specific intervals.

Going Further

eLinux's Raspberry Pi GPIO Reference Page (http://bit.ly/ 1vORWl3)

> This is the most comprehensive reference guide to the Raspberry Pi's GPIO pins.

Gordon Henderson's Command Line GPIO Utility (https:// projects.drogon.net/raspberry-pi/wiringpi/the-gpio-utility)

> This command line utility makes it easier to work with GPIO pins from the command line. It's bundled with the latest version of Raspbian. Try running the command `gpio readall` to get an overview of all your pins.

Adafruit: MCP230xx GPIO Expander on the Raspberry Pi (http:// bit.ly/1oTYdGT)

> If you don't have enough pins to work with, Adafruit offers this guide to using the MCP23008 chip for 8 extra GPIO pins and the MCP23017 for 16 extra GPIO pins.

7/Programming Inputs and Outputs with Python

At the end of Chapter 6, you did a little bit of programming with the Raspberry Pi's GPIO pins using a shell script. In this chapter, you're going to learn how to use Python to do the same thing...and a little more. As with the shell script, Python will let you access the GPIO pins with code that reads and controls the pins automatically.

The advantage that Python has over shell scripting is that Python code is easier to write and is more readable. There's also a whole slew of Python modules that make it easy for you to do some complex stuff with basic code. See Table 4-2 for a list of a few modules that might be useful in your projects. Best of all, there's a Python module called RPi.GPIO (*http://bit.ly/1vzTBtl*) that makes it easy to read and control the GPIO pins. You're going to learn how to use that module in this chapter.

Testing GPIO in Python

1. Go into the Python interactive interpreter from the terminal prompt. (In older versions of Raspbian, *RPi.GPIO* requires root access to read and control the pins. If you're using an older version of Raspbian and get a warning about permis-

sions, you'll need to launch Python as root with the command `sudo python` when using RPi.GPIO.)

Here's how to do it:

```
$ python
Python 2.7.3rc2 (default, May  6 2012, 20:02:25)
[GCC 4.6.3] on linux2
Type "help", "copyright", "credits" or "license" for more
    information.
>>>
```

2. When you're at the >>> prompt, try importing the module:

```
>>> import RPi.GPIO as GPIO
```

In this chapter, we'll be using Python 2.7 instead of Python 3 because one of the modules we'll be using is only installed for Python 2.x on the Raspberry Pi. When you type **python** at the command prompt on the Raspberry Pi, it runs Python 2.7 by default. This behavior could change in the future (you can run Python 2.7 explicitly by typing **python2.7** instead of **python**).

One important difference between the two versions is how you print text to the console. You'll use `print "Hello, world!"` in Python 2.x, but you'd use `print("Hello, world!")` in Python 3.

If you don't get any errors after entering the import command, you know you're ready to try it out for the first time:

1. Before you can use the pins, you must tell the GPIO module how your code will refer to them. In Chapter 6, the pin numbers we used didn't correlate to the way that they're arranged on the board. You were actually using the on-board Broadcom chip's signal name for each pin. With this Python module, you can choose to refer to the pins either way. To use the numbering from the physical layout, use `GPIO.set mode(GPIO.BOARD)`. But let's stick with the pin numbering that you used in Chapter 6 (`GPIO.setmode(GPIO.BCM)`, where BCM

is short for Broadcom), which is what Adafruit's Pi Cobbler and similar breakout boards use for labels:

```
>>> GPIO.setmode(GPIO.BCM)
```

2. Set the direction of pin 25 to output:

```
>>> GPIO.setup(25, GPIO.OUT)
```

3. Connect an LED to pin 25 as you did in "Beginner's Guide to Breadboarding" on page 84.

4. Turn on the LED:

```
>>> GPIO.output(25, GPIO.HIGH)
```

5. Turn off the LED:

```
>>> GPIO.output(25, GPIO.LOW)
```

6. Unexport the pins that were used in this session:

```
>>> GPIO.cleanup()
```

7. Exit the Python interactive interpreter:

```
>>> exit()
$
```

In Chapter 6, you learned that digital input and output signals on the Raspberry Pi must be either 3.3 volts or ground. In digital electronics, we refer to these signals as high or low, respectively. Keep in mind that not all hardware out there uses 3.3 volts to indicate high; some use 1.8 volts or 5 volts. If you plan on connecting your Raspberry Pi to digital hardware through its GPIO pins, it's important that they also use 3.3 volts.

Those steps gave you a rough idea of how to control the GPIO pins by typing Python statements directly into the interactive interpreter. Just as you created a shell script to turn the pins on and off in Chapter 6, you're going to create a Python script to read and control the pins automatically.

Blinking an LED

To blink an LED on and off with Python, you're going to use the statements that you already tried in the interactive interpreter in addition to a few others. For the next few steps, we'll assume you're using the desktop environment (as shown in Figure 7-1), but feel free to use the command line to write and execute these Python scripts if you prefer.

Here's how to blink an LED with a Python script:

1. Open the Leafpad text editor by clicking the Raspberry Menu→ Accessories→Text Editor.

2. Enter the following code:

```
import RPi.GPIO as GPIO  ❶
import time  ❷

GPIO.setmode(GPIO.BCM)  ❸
GPIO.setup(25, GPIO.OUT)  ❹

while True:  ❺
    GPIO.output(25, GPIO.HIGH)  ❻
    time.sleep(1)  ❼
    GPIO.output(25, GPIO.LOW)  ❽
    time.sleep(1)  ❾
```

❶ Import the code needed for GPIO control.

❷ Import the code needed for for the sleep function.

❸ Use the chip's signal numbers.

❹ Set pin 25 as an output.

❺ Create an infinite loop consisting of the indented code below it.

❻ Turn the LED on.

❼ Wait for one second.

❽ Turn the LED off.

❾ Wait for one second.

3. Save the file as *blink.py* within the home directory, */home/pi*. There's a shortcut to this folder in the places list on the left side of the Save As window.

4. Open LXTerminal, then use these commands to make sure the working directory is your home directory, and execute the script (see Figure 7-1):

```
pi@raspberrypi ~/Documents $ cd ~
pi@raspberrypi ~ $ python blink.py
```

5. Your LED should now be blinking!

6. Press Ctrl-C to stop the script and return to the command line.

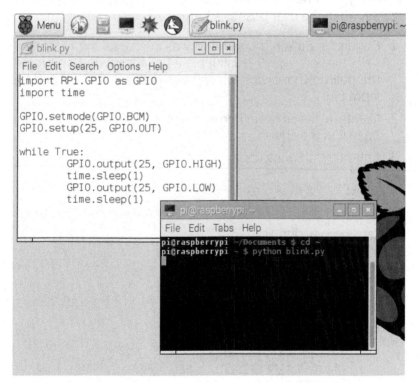

Figure 7-1. *Using Leafpad and LXTerminal to edit and launch Python scripts.*

Try modifying the script to make the LED blink faster by using decimals in the `time.sleep()` functions. You can also try adding a few more LEDs and getting them to blink in a pattern. You can use any of the dedicated GPIO pins that are shown in Figure 6-2.

Reading a Button

If you want something to happen when you press a button, one way to do that is to use a technique called *polling*. Polling means continually checking over and over again for some condition. In this case, your program will be polling whether the button is connecting the input pin to 3.3 volts or to ground. To learn about polling, you'll create a new Python script that will display text on screen when the user pushes a button:

1. Connect a button the same way as in "Digital Input: Reading a Button" on page 89, using pin 24 as the input. Don't forget the pull-down resistor, which goes between ground and the input pin.

2. Create a file in your home directory called *button.py* and open it in the editor.

3. Enter the following code:

```
import RPi.GPIO as GPIO
import time

GPIO.setmode(GPIO.BCM)
GPIO.setup(24, GPIO.IN)    ❶

count = 0  ❷

while True:
    inputValue = GPIO.input(24)    ❸
    if (inputValue == True):    ❹
        count = count + 1    ❺
        print("Button pressed " + str(count) + " times.")    ❻
    time.sleep(.01)    ❼
```

❶   Set pin 24 as an input.

❷   Create a variable called **count** and store 0 in it.

❸   Save the value of pin 24 into **inputValue**.

❹ Check if that value is **True** (when the button is pressed).

❺ If it is, increment the counter.

❻ Print the text to the terminal.

❼ Wait briefly, but let other programs have a chance to run by not hogging the processor's time.

4. Go back to LXTerminal and execute the script:

    ```
    $ python button.py
    ```

5. Now press the button. If you've got everything right, you'll see a few lines of "The button has been pressed" for each time you press the button.

The preceding code checks for the status of the button 100 times per second, which is why you'll see more than one sentence printed (unless you have incredibly fast fingers). The Python statement `time.sleep(.01)` is what controls how often the button is checked.

But why not continually check the button? If you were to remove the `time.sleep(.01)` statement from the program, the loop would indeed run incredibly fast, so you'd know much more quickly when the button was pressed. This comes with a few drawbacks: you'd be using the processor on the board constantly, which will make it difficult for other programs to function, and it would increase the Raspberry Pi's power consumption. Because *button.py* has to share resources with other programs, you have to be careful that it doesn't hog them all up.

Now add a few lines to the code to make it a little bit better at registering a single button press:

```python
import RPi.GPIO as GPIO
import time

GPIO.setmode(GPIO.BCM)
GPIO.setup(24, GPIO.IN)

count = 0

while True:
    inputValue = GPIO.input(24)
```

```
if (inputValue == True):
    count = count + 1
    print("Button pressed " + str(count) + " times.")
    time.sleep(.3)  ❶
time.sleep(.01)
```

❶   Helps a button press register only once.

This additional line of code will help to ensure that each button press is registered only once. But it's not a perfect solution. Try holding down the button. Another count will be registered and displayed three times a second, even though you're holding the button down. Try pushing the button repeatedly very quickly. It doesn't register every button press because it won't recognize distinct button presses more frequently than three times a second.

These are challenges that you'll face when you're using polling to check the status of a digital input. One way to get around these challenges is to use an *interrupt*, which is a way of setting a specified block of code to run when the hardware senses a change in the state of the pin. RPi.GPIO supports interrupts and you can read about how to use this feature in the library's documentation (*http://pypi.python.org/pypi/RPi.GPIO*).

Project: Simple Soundboard

Now that you know how to read the inputs on the Raspberry Pi, you can use the Python module Pygame's sound functions to make a soundboard. A soundboard lets you play small sound recordings when you push its buttons. To make your own soundboard, you'll need the following in addition to your Raspberry Pi:

- Three pushbutton switches
- Female-to-male jumper wires
- Standard jumper wires or hookup wire, cut to size
- Solderless breadboard
- Three resistors, 10K ohm
- Computer speakers, or an HDMI monitor that has built-in speakers

You'll also need a few uncompressed sound files, in *.wav* format. For purposes of testing, there are a few sound files preloaded on the Raspberry Pi that you can use. Once you get the soundboard working, it's easy to replace those files with any sounds you want, though you may have to convert them to *.wav* from other formats. Start off by building the circuit:

1. Using a female-to-male jumper wire, connect the Raspberry Pi's ground pin to the negative rail on the breadboard.

2. Using a female-to-male jumper wire, connect the Raspberry Pi's 3.3V pin to the positive rail on the breadboard.

3. Insert the three pushbutton switches in the breadboard, all straddling the center trench.

4. Using standard jumper wires or small pieces of hookup wire, connect the positive rail of the breadboard to the top pin of each button.

5. Now add the pull-down resistors. Connect the bottom pin of each button to ground with a 10K resistor.

6. Using female-to-male jumper wires, connect each button's bottom pin (the one with the 10K resistor) to the Raspberry Pi's GPIO pins. For this project, we used pins 23, 24, and 25.

Figure 7-2 shows the completed circuit. We created this diagram with Fritzing (*http://fritzing.org*), an open source tool for creating hardware designs.

Figure 7-2. *Completed circuit for the soundboard project*

Now that you have the circuit breadboarded, it's time to work on the code:

1. Create a new directory in your home directory called *soundboard*.

2. Open that folder and create a file there called *soundboard.py*.

3. Open *soundboard.py* and type in the following code:

```python
import pygame.mixer
from time import sleep
import RPi.GPIO as GPIO
from sys import exit

GPIO.setmode(GPIO.BCM)
GPIO.setup(23, GPIO.IN)
GPIO.setup(24, GPIO.IN)
GPIO.setup(25, GPIO.IN)

pygame.mixer.init(48000, -16, 1, 1024)   ❶

soundA = pygame.mixer.Sound(
    "/usr/share/sounds/alsa/Front_Center.wav")   ❷
soundB = pygame.mixer.Sound(
    "/usr/share/sounds/alsa/Front_Left.wav")
soundC = pygame.mixer.Sound(
    "/usr/share/sounds/alsa/Front_Right.wav")

soundChannelA = pygame.mixer.Channel(1)   ❸
soundChannelB = pygame.mixer.Channel(2)
soundChannelC = pygame.mixer.Channel(3)

print "Soundboard Ready."   ❹

while True:
    try:
        if (GPIO.input(23) == True):   ❺
            soundChannelA.play(soundA)   ❻
        if (GPIO.input(24) == True):
            soundChannelB.play(soundB)
        if (GPIO.input(25) == True):
            soundChannelC.play(soundC)
        sleep(.01)   ❼
```

```
    except KeyboardInterrupt:  ⑧
        exit()
```

❶ Initialize Pygame's mixer.

❷ Load the sounds.

❸ Set up three channels (one for each sound) so that we can play different sounds concurrently.

❹ Let the user know the soundboard is ready (using Python 2 syntax).

❺ If the pin is high, execute the following line.

❻ Play the sound.

❼ Don't "peg" the processor by checking the buttons faster than we need to.

❽ This will let us exit the script cleanly when the user hits Ctrl-C, without showing the traceback message.

4. Go to the command line and navigate to the folder where you've saved *soundboard.py* and execute the script with Python 2:

```
pi@raspberrypi ~/soundboard $ python soundboard.py
```

5. After you see "Soundboard Ready," start pushing the buttons to play the sound samples.

While Pygame is available for Python 3, on the Raspberry Pi's default installation, it's only installed for Python 2.

Depending on how your Raspberry Pi is set up, your sound might be sent via HDMI to your display, or it may be sent to the 3.5mm analog audio output jack on the board. To change that, exit out of the script by pressing Ctrl-C and executing the following command to use the analog audio output:

```
pi@raspberrypi ~/soundboard $ sudo amixer cset numid=3 1
```

To send the audio through HDMI to the monitor, use:

```
pi@raspberrypi ~/soundboard $ sudo amixer cset numid=3 2
```

Of course, the stock sounds aren't very interesting, but you can replace them with any of your own sounds: applause, laughter, buzzers, and dings. Add them to the *soundboard* directory and update the code to point to those files. If you want to use more sounds on your soundboard, add additional buttons and update the code as necessary.

Going Further

RPi.GPIO (https://sourceforge.net/projects/raspberry-gpio-python)
> Because the RPi.GPIO library is still under active development, you may want to check the homepage for the project to find the latest updates.

8/Analog Input and Output

In Chapter 6, you learned about digital inputs and outputs with buttons, switches, LEDs, and relays. Each of these components was always either on or off, never anything in between. However, you might want to sense things in the world that are not necessarily one or the other —for instance, temperature, distance, light levels, or the status of a dial. These all come in a range of values.

Or you may want to put something in a state that's not just on or off. For example, if you wanted to dim an LED or control the speed of a motor rather than just turning it on or off. To make an analogy for analog and digital, you can think of a typical light switch versus a dimmer switch, as pictured in Figure 8-1.

Figure 8-1. *Digital is like the switch on the left: it can be either on or off. Analog, on the other hand, can be set at a range of values between fully on and completely off.*

Output: Converting Digital to Analog

Just as in Chapter 7, you'll use the GPIO Python module already installed in the most recent versions of Raspbian Linux. The module has experimental functions for controlling the GPIO pins, sort of like a dimmer switch.

We say that it's "sort of" like a dimmer switch because the module uses a method called *pulse-width modulation*, or PWM, to make it *seem* like there's a range of voltages coming out of its outputs. What it's actually doing is pulsing its pins on and off really quickly. So if you want the pin to behave as though it's at half voltage, the pin will be pulsed so that it is on 50% of the time and off 50% of the time. If you want the pin to behave as though it's at 20% power, the Raspberry Pi will turn the pin on 20% of the time and off 80% of the time. The percentage of time that it's on versus total time of the cycle is called the *duty cycle* (Figure 8-2). When you connect an LED to these pins and instruct the Raspberry Pi to change the duty cycle, it has the effect of dimming the LED.

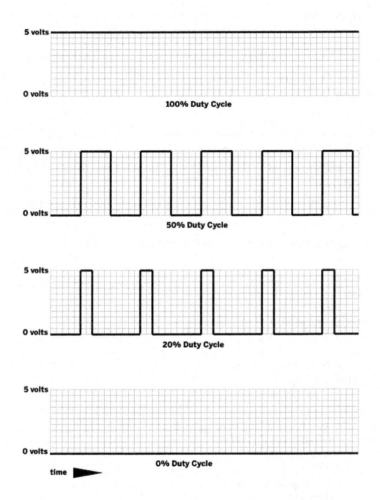

Figure 8-2. *The duty cycle represents how much time the pin is turned on over the course of an on-off cycle*

Test-Driving PWM

For the next few steps, we'll assume you're using the desktop environment, but feel free to use the command line to write and execute these Python scripts if you prefer:

1. Connect an LED to pin 25 as you did in "Beginner's Guide to Breadboarding" on page 84.

2. Open the File Manager by clicking its icon in the taskbar.

3. Be sure you're in the home directory, the default being /home/pi. If not, click on the home icon under the Places listing.

4. Create a file in your home directory called *blink.py*. Do this by right-clicking in the home directory window, going to "Create New" and then clicking "Blank File." Name the file *fade.py*.

5. Double-click on *fade.py* to open it in Leafpad, the default text editor.

6. Enter the following code and save the file:

```
import RPi.GPIO as GPIO
import time

GPIO.setmode(GPIO.BCM)
GPIO.setup(25, GPIO.OUT)

p = GPIO.PWM(25, 50)  ❶
p.start(0)  ❷

while True:
    for dc in range(0, 100, 5):  ❸
        p.ChangeDutyCycle(dc)  ❹
        time.sleep(0.05)
    for dc in range(100, 0, -5):  ❺
        p.ChangeDutyCycle(dc)
        time.sleep(0.05)
```

❶ Create a PWM object called **p** and set it to GPIO pin 25 with a frequency of 50 Hz.

❷ Start PWM on **p**.

❸ Run the indented code below this line, each time incrementing the value of dc by 5 from starting at 0 and going to 100.

❹ Set the duty cycle of p to the value of dc.

❺ Run the indented code that follows, each time decrementing the value of dc by 5 from starting at 100 and going to 0.

7. Open LXTerminal, then use these commands to make sure the working directory is your home directory, and execute the script:

```
pi@raspberrypi ~/Development $ cd ~
pi@raspberrypi ~ $ python fade.py
```

8. Your LED should now be fading up and down!

9. Press Ctrl-C to stop the script and return to the command line.

If you're accustomed to using PWM on a microcontroller like the Arduino, you'll find that—unlike Arduino—there is unsteadiness in the PWM pulses from the Raspberry Pi. This is because in this example, you're using the CPU to turn the LED on and off. Because that CPU is used for multiple things at one time, it may not always keep perfect time. You can always reach for other hardware like Adafruit's PWM/Servo Driver (*https://www.adafruit.com/products/815*) if you need to have more precise control.

Taking PWM Further

With the ability to use pulse-width modulation to fade LEDs up and down, you could also connect an RGB LED and control the color by changing the brightness of its red, green, and blue elements.

Pulse-width modulation can also be used to control the speed of a direct current motor that's connected to your Raspberry Pi through transistors. The position of the shaft on a hobby servo motor (the kind that steer RC cars) can also be controlled with specific pulses of electricity. Though, keep in mind that you may need additional hardware and power in order to control these motors with a Raspberry Pi.

Input: Converting Analog to Digital

Just like you controlled the output of a GPIO pin on a scale of 0 to 100, it's also possible to read sensors that can offer the Raspberry Pi a range of values. If you want to know the temperature of a room, the light level, or the amount of pressure on a pad, you can use various sensors. On a microcontroller like the Arduino, there's special hardware to convert the analog voltage level to digital data. Unfortunately, your Raspberry Pi doesn't have have this hardware built in.

To convert from analog to digital, this section will show you how to use an *ADC*, or *analog-to-digital converter*. There are a few different models of ADCs out there, but this chapter will use the ADS1x15 from Texas Instruments. The package of the ADS1x15 chip is too small for your standard breadboard, so Adafruit Industries has created a breakout board (*http:// www.adafruit.com/products/1083*) for it, shown in Figure 8-3. Once you've soldered header pins to the breakout board, you can prototype with this chip in your breadboard. The chip uses a protocol named *Inter-Integrated Circuit* (commonly called *I2C*) for transmitting the analog readings. Luckily, we don't need to understand the protocol in order to use it. Adafruit provides an excellent open source Python library to read the values from the ADS1115 or its little brother, the ADS1015, via I2C.

To connect the ADS1115 or ADS1015 breakout to your Raspberry Pi:

1. Connect the 3.3 volt pin from the Raspberry Pi to the positive rail of the breadboard. Refer to Figure 6-2 for pin locations on the Raspberry Pi's GPIO header.

2. Connect the ground pin from the Raspberry Pi to the negative rail of the breadboard.

3. Insert the ADS1x15 into the breadboard and use jumper wires to connect its VDD pin to the positive rail and its GND pin to the negative rail.

4. Connect the SCL pin on the ADS1x15 to the SCL pin on the Raspberry Pi. The SCL pin on the Pi is the one paired with the ground pin on the GPIO header.

5. Connect the SDA pin on the ADS1x15 to the SDA pin on the Raspberry Pi. The SDA pin is in between the the SCL pin and the 3.3 volt pin.

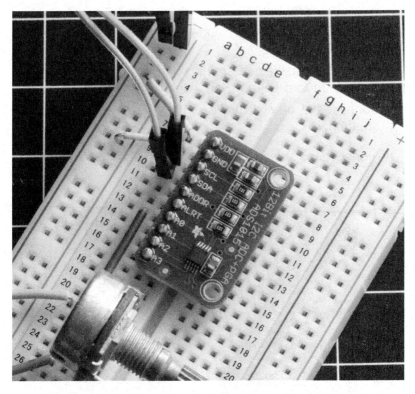

Figure 8-3. *The ADS1015 analog-to-digital converter breakout from Adafruit*

Now you'll need to connect an analog sensor to the ADS1x15. There are many to choose from, but for this walk-through, you'll use a 2K potentiometer so that we can have a dial input for your

Raspberry Pi. A *potentiometer*, or *pot*, is essentially a variable voltage divider, and can come in the form of a dial or slider.

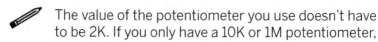 The value of the potentiometer you use doesn't have to be 2K. If you only have a 10K or 1M potentiometer, it will work just fine.

To connect a potentiometer to the ADS1x15:

1. Insert the potentiometer into the breadboard.
2. The pot has three pins. Connect the middle pin to pin A0 in on the ADS1x15.
3. One of the outside pins should connect to the positive rail of the breadboard. For now, it doesn't matter which.
4. Connect the other outside pin to the negative rail of the breadboard.

The connections should look as shown in Figure 8-4.

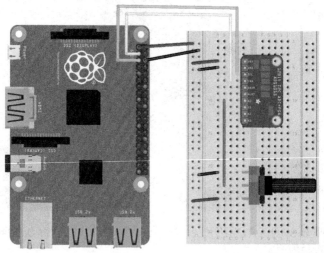

fritzing

Figure 8-4. *Using the ADS1x15 to connect a potentiometer to the Raspberry Pi*

Before you can read the potentiometer, you'll need to enable I2C and install a a few things (see Figure 8-5):

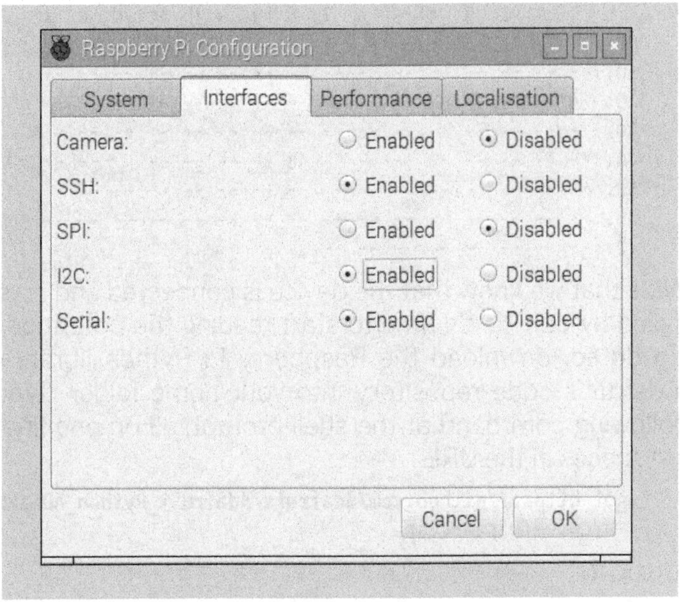

Figure 8-5. *Enabling the I2C interface*

1. Open up the Raspberry Pi Configuration tool by clicking on Menu→Preferences→Raspberry Pi Configuration.
2. Click the Interfaces tab.
3. Next to I2C, click Enable.
4. Click OK and reboot your Raspberry Pi.
5. Open up a Terminal window and update your list of packages:

```
$ sudo apt-get update
```

6. Install i2c-tools tools:

```
$ sudo apt-get install i2c-tools
```

7. Restart your Raspberry Pi.
8. After you've restarted your Raspberry Pi, test that the Raspberry Pi can detect the ADS1x15 with the command:

```
$ sudo i2cdetect -y 1
```

9. If the board is recognized, you'll see the number in the grid that is displayed:

```
     0  1  2  3  4  5  6  7  8  9  a  b  c  d  e  f
00:           -- -- -- -- -- -- -- -- -- -- -- -- --
10: -- -- -- -- -- -- -- -- -- -- -- -- -- -- -- --
20: -- -- -- -- -- -- -- -- -- -- -- -- -- -- -- --
30: -- -- -- -- -- -- -- -- -- -- -- -- -- -- -- --
40: -- -- -- -- -- -- -- -- 48 -- -- -- -- -- -- --
50: -- -- -- -- -- -- -- -- -- -- -- -- -- -- -- --
60: -- -- -- -- -- -- -- -- -- -- -- -- -- -- -- --
70: -- -- -- -- -- -- -- --
```

10. Now that we know that the device is connected and is recognized by our Pi, it's time to start reading the potentiometer. To do so, download the Raspberry Pi Python library from Adafruit's code repository into your home folder. Type the following command at the shell prompt, all on one line with no spaces in the URL:

    ```
    wget https://github.com/adafruit/Adafruit_Python_ADS1x15/
        archive/master.zip
    ```

11. Unzip it:

    ```
    $ unzip master.zip
    ```

12. Change to the library's ADS1x15 directory:

    ```
    $ cd Adafruit_Python_ADS1x15-master/
    ```

13. Install the library:

    ```
    $ sudo python setup.py install/
    ```

14. Run one of the example files:

    ```
    $ python examples/simpletest.py
    ```

15. Turn the potentiometer all the way in one direction and back in the other. Notice how the value in the pin 0 column changes:

```
Reading ADS1x15 values, press Ctrl-C to quit...
|     0 |     1 |     2 |     3 |
-------------------------------------
|     0 |  4320 |  4496 |  4528 |
|     0 |  4224 |  5104 |  4432 |
|     0 |  4352 |  4688 |  5104 |
|  6128 |  4640 |  4384 |  5552 |
| 10928 |  4592 |  4592 |  5056 |
| 18736 |  4384 |  4992 |  4384 |
```

23312	4496	4912	4784
26512	4912	4800	5968
25968	4816	4800	5008
16528	4416	4928	4928
9280	4656	4416	5312
2688	4240	5008	4464
0	4384	4352	5360
0	4336	4624	4272
0	4256	4432	4592

16. Press Ctrl-C to terminate the script.

As you can see, turning the dial on the potentiometer changes the voltage coming into channel 0 of the ADS1x15. The code in the example does a little bit of math to determine the voltage value from the data coming from the ADC. Of course, your math will vary depending on what kind of sensor you use.

You can look inside that example script to see how it works, or try out the even simpler example that follows to learn how to take readings from the ADC. Create a new file with the code from Example 8-1 and execute it.

Example 8-1. Writing the code to read the ADC

```
from Adafruit_ADS1x15 ❶
from time import sleep

adc = Adafruit_ADS1x15.ADS1115() ❷

while True:
    result = adc.read_adc(0) ❸
    print result
    sleep(.5)
```

❶ Import Adafruit's ADS1x15 library.

❷ Create a new ADS1x15 object called adc.

❸ Get a reading from channel A0 on the ADS1x15 and store it in result.

When you run this code, it will output out raw numbers for each reading twice a second. Turning the potentiometer will make the values go up or down.

Once you get it all set up, the Adafruit ADS1x15 library does all the hard work for you and makes it easy to use analog sensors in your projects. For instance, if you want to make your own Pong-like game, you could read two potentiometers and then use Pygame to draw on the game on screen. For more information about using Pygame, see Chapter 4.

Variable Resistors

Not all analog sensors work like the potentiometer, which provides a variable voltage based on some factor (such as the amount the dial on the pot is turned).

Some sensors are simply variable resistors that resist the flow of electricity based on some factor. For instance, a photocell like the one in Figure 8-6 will act like a resistor that changes values based on the amount of light hitting the cell. Add more light, and the resistance goes down. Take away light, and the resistance goes up. A force-sensitive resistor decreases its resistance as you put pressure on the pad.

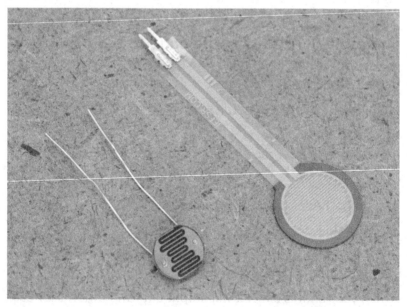

Figure 8-6. *The photocell and force-sensitive resistors act as variable resistors and can be used as analog inputs*

In order to read sensors like these with analog input pins, you'll need to incorporate a *voltage divider* circuit.

Voltage Divider Circuit

When you're working with sensors that offer variable resistance, the purpose of a voltage divider is to convert the variable resistance into a variable voltage, which is what the analog input pins are measuring. First, take a look at a simple voltage divider.

In Figure 8-7, you'll see two resistors of the same value in series between the positive and ground and one wire connected to analog input 0 between the two resistors. Because both resistors are 10K ohms, there's about 1.65 volts going to analog input 0, half of 3.3 volts.

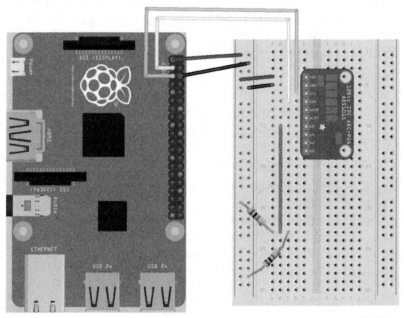

fritzing

Figure 8-7. *With two of the same value resistors between voltage and ground, the voltage between the two would be half of the total voltage*

Without getting bogged down in the math involved, if you removed the 10K resistor connected to 3.3 volts and replaced it with a resistor of a higher value, the voltage going into the analog pin would *decrease*. If you removed that 10K resistor and replaced it with one of a lower value, the voltage going into the analog pin would *increase*. We can use this principle with sensors that are variable resistors in order to read them with the analog input pins. You'll simply replace one of the resistors with your sensor.

To try the circuit out, you'll have to wire up a type of variable resistor called a *force-sensitive resistor*, or FSR.

Force-Sensitive Resistor

A force-sensitive resistor is a variable resistor that changes based on the amount of pressure placed on its pad. When there's no pressure on the pad, the circuit is open. When you start placing pressure on it, the resistance drops.

The exact figures will depend on your particular FSR, but typically you'll see 100K ohms of resistance with light pressure and 1 ohm of resistance with maximum pressure. If you have a *multimeter*, you can measure the changes in resistance to see for yourself, or you can look at the component's *datasheet*, which will tell you what to expect from the sensor.

If you're going to replace the resistor connected to 3.3 volts in Figure 8-7 with a variable resistor like an FSR, you'll want the value of the other resistor to be somewhere in between the minimum and maximum resistance so that you can get the most range out of the sensor. For a typical FSR, try a 10K ohm resistor. To give the force-sensitive resistor a test drive:

1. Wire up an FSR to the ADC as shown in Figure 8-8.
2. Execute the code in Example 8-1 again.
3. Watch the readings on the screen as you squeeze the FSR.

If everything is working correctly, you should see the values rise as you increasingly squeeze harder on the FSR. As you press harder, you're reducing the resistance between the two leads on the FSR. This has the effect of sending higher voltage to the analog input.

You'll encounter many analog sensors that use this very same principle, and a simple voltage divider circuit, along with an analog-to-digital converter, will allow your Raspberry Pi to capture that sensor data.

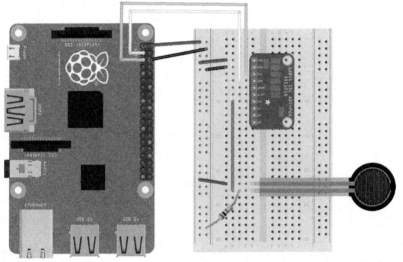

fritzing

Figure 8-8. *Wiring up a force-sensitive resistor to the ADC requires a voltage divider circuit*

Going Further

Controlling a DC Motor with PWM (http://bit.ly/1xpmBlF)
Adafruit has an excellent guide that teaches you how to use PWM to control the speed of a DC motor.

Reading Resistive Sensors with RC Timing (http://bit.ly/1rk6yJl)
Adafruit also shows you how to use a simple circuit to read resistive analog sensors without an analog-to-digital converter.

9/Working with Cameras

One of the advantages to using a platform like the Raspberry Pi for DIY technology projects is that it supports a wide range of peripheral devices. Not only can you hook up a keyboard and mouse, you can also connect accessories like printers, WiFi adapters (at least on the first two generations of Raspberry Pi—the Model 3 has its own built in), thumb drives, additional memory cards, cameras, and hard drives. In this chapter, we're going to show you a few ways to use a camera in your Raspberry Pi projects.

While not quite as common as a keyboard and mouse, a webcam is almost a standard peripheral for computers these days. Most laptops sold have a tiny camera built into the bezel of the display. And if they don't have a built-in camera, a USB webcam from a well-known brand can be purchased for as little as $25. You can even find webcams for much less if you take a chance on an unknown brand.

The folks at the Raspberry Pi Foundation have created their own camera peripheral that is designed to work with Raspberry Pi (Figure 9-1). Unlike a USB webcam, you're unlikely to find the official Raspberry Pi camera module in an office supply store, but you should be able to buy it wherever Raspberry Pis are sold, for around $25.

Figure 9-1. *Raspberry Pi's camera module*

Unlike a USB webcam, the camera board connects to Raspberry Pi's Camera Serial Interface (CSI) connector (Figure 9-2). The reason is this: since the Broadcom chip at the core of the Raspberry Pi is meant for mobile phones and tablets, the CSI connection is how a mobile device manufacturer would connect a camera to the chip. Throughout this chapter, we'll use the official camera board as our chief example, but many of the projects and tutorials can also be done with a USB webcam (Figure 9-3).

Figure 9-2. *Raspberry Pi's camera serial interface*

Figure 9-3. *A typical USB webcam*

Connecting and Testing the Camera Module

Connecting the official camera module isn't as straightforward as connecting a USB device, but once you get it working, it should be a piece of cake.

 Make sure the Raspberry Pi is powered down before you do this.

Here are the steps you'll need to take:

1. Pull up on edges of the CSI connector, which is right next to the Ethernet port. A piece of the connector will slide up and lean back toward the Ethernet port. (See Figure 9-4.)

Figure 9-4. *Opening the camera serial interface connector locking mechanism*

2. Insert the camera module's ribbon cable into the CSI connector so that its metal contacts are facing away from the Ethernet port.

3. Hold the ribbon cable into the CSI connector and press the moving part of the CSI connector down to lock and hold the ribbon cable in place. You'll still see part of the metal contacts on the ribbon cable. (See Figure 9-5.)

Figure 9-5. *After placing the ribbon cable into the camera serial interface connector and locking it down, you may still see metal contacts on the ribbon cable*

4. Open the Raspberry Pi Configuration tool from the desktop Menu→Preferences. (See Figure 9-6)

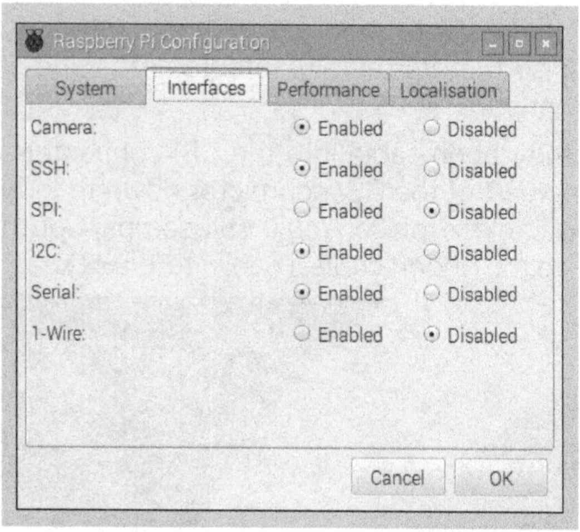

Figure 9-6. *Enabling the camera interface in the Raspberry Pi configuration tool.*

5. Click the "Interfaces" tab.
6. Click the option to enable the camera.
7. Click OK and reboot your Raspberry Pi.
8. After you reboot, open a Terminal session and test the camera with:

```
$ raspistill -o test.jpg
```

If everything works, you'll see a preview image from the camera appear on screen for a few seconds. After it disappears, a JPEG file of the captured image will be saved in your current directory.

raspistill is a powerful program with a lot of options. To see what's possible with it, view all the command-line options by running the program and piping the output through less:

```
$ raspistill 2>&1 | less
```

Use the up and down arrow keys to scroll through the options, and press q when you want to get back to the command line.

Project: Making a GIF

One of the features of raspistill is that it can take a series of photos at a specific interval. We can use this feature, along with the command-line image converting and editing software *ImageMagick*, to create fun animated GIFs with the Raspberry Pi:

1. First, install ImageMagick:

   ```
   $ sudo apt-get update
   $ sudo apt-get install imagemagick
   ```

2. Create a new directory to hold the images you capture and switch to that directory:

   ```
   $ mkdir gif-test
   $ cd gif-test
   ```

3. With your camera ready, execute `raspistill` to run for nine seconds, taking a 640 × 480 resolution image every three seconds, naming each file with an incrementing filename:

   ```
   $ raspistill -t 9000 -tl 3000 -o image%02d.jpg -w
     640 -h 480
   ```

4. Next, input those files into ImageMagick outputting as the file *test.gif*:

   ```
   $ convert -delay 15 *.jpg test.gif
   ```

5. Now open the *test.gif* by double-clicking the file within the desktop environment, and you'll see the GIF you made!

Capturing Video

There's also a command-line utility called `raspivid` to capture video from the official Raspberry Pi camera module. Try capturing a five-second video and saving it to a file:

```
$ raspivid -t 5000 -o video.h264
```

You can play that file back with:

```
$ omxplayer video.h264
```

And just like `raspistill`, `raspivid` is a powerful program with a lot of options. To see what's possible with it, view all the command-line options by running the program and piping the output through `less`:

```
$ raspivid 2>&1 less
```

Testing USB Webcams

With all the different models of webcams out there, there's no guarantee that a camera will work right out of the box. If you're purchasing a webcam for use with the Raspberry Pi, search online to make sure that others have had success with the model that you're purchasing. You can also check the webcam section of eLinux.org's page of peripherals (*http://elinux.org/ RPi_VerifiedPeripherals#USB_Webcams*) that have been verified to work with the Raspberry Pi.

Be aware that you'll need to connect a powered USB hub to your Raspberry Pi if you want to connect your webcam in addition to your keyboard and mouse. The hub must be powered because the Raspberry Pi only lets a limited amount of electrical current through its USB ports, and it likely won't be able to provide enough power for your keyboard, mouse, *and* webcam. A powered USB hub plugs into the wall and provides electrical current to the peripherals that connect to it so that they don't max out the power on your Raspberry Pi.

If you have a webcam that you're ready to test out with the Raspberry Pi, use `apt-get` in the Terminal to install a simple camera viewing application called luvcview:

```
$ sudo apt-get install luvcview
```

After `apt-get` finishes the installation, run the application by typing `luvcview` in a Terminal window while you're in the desktop environment. A window will open showing the view of the first video source it finds in the */dev* folder, likely */dev/video0*. Note the frame size that is printed in the Terminal window. If the video seems a little choppy, you can fix this by reducing the default size of the video. For example, if the default video size is 640 × 480, close luvcview and reopen it at half the video size by typing the following at the command line:

```
$ luvcview -s 320x240
```

If you don't see video coming through, you'll want to troubleshoot here before moving on. One way to see what's wrong is by disconnecting the webcam, reconnecting it, and running the command `dmesg`, which will output diagnostic messages that might give you some clues as to what's wrong.

Installing and Testing SimpleCV

In order to access the camera with Python code, we'll use SimpleCV (Figure 9-7), which is a feature-packed open source computer vision library. SimpleCV makes it really easy to get images from the camera, display them on screen, or save them as files. But what makes SimpleCV really stand out are its computer vision algorithms, which can do some pretty amazing things. Besides basic image transformations, it can also track, detect, and recognize objects in an image or video. Later on in this chapter, we'll try basic face detection with SimpleCV ("Face Detection" on page 145).

Figure 9-7. *SimpleCV logo*

To install SimpleCV for Python, you'll need to start by installing the other libraries it depends on. For those, you can use `apt-get`:

```
$ sudo apt-get install python-opencv python-scipy python-numpy
  python-pip
$ sudo pip install svgwrite
```

It's a big install, and it may take a while before the process is complete. Next, you'll install the actual SimpleCV library with the following command:

```
$ sudo pip install https://github.com/sightmachine/SimpleCV/
  zipball/master
```

When it's done, check that the installation worked by going into the Python interactive interpreter and importing the library:

```
$ python
Python 2.7.3rc2 (default, May  6 2012, 20:02:25)
[GCC 4.6.3] on linux2
Type "help", "copyright", "credits" or "license" for more
  information.
```

```
>>> import SimpleCV
>>>
```

If you get no errors after importing the library, you know you've got SimpleCV installed correctly. If you're using a USB webcam, you can jump ahead to "Displaying an Image" on page 138. If you're using the Raspberry Pi camera module, there are a few extra steps, which we'll cover now.

Additional Steps for the Raspberry Pi Camera Module

Because SimpleCV uses something called the Video4Linux2 (V4L2) driver framework to access USB webcams, we'll need to load an extra kernel driver so that the official camera board can be used with SimpleCV. If you're using a regular USB webcam, you can skip these steps:

1. From the Terminal, run the following command to edit the configuration file:

   $ `sudo nano /etc/modules-load.d/modules.conf`

2. Add the following line to the end of that file:

   `bcm2835-v4l2`

3. Exit nano by pressing Ctrl-X followed by y to save the file when it prompts you.

4. After you reboot your Raspberry Pi, the V4L2 kernel driver for the Raspberry Pi Camera Module will be loaded.

Displaying an Image

For many of the examples in this chapter, you'll need to work in the desktop environment so that you can display images on screen. You can work in IDLE, or save your code as .py files from Leafpad and execute them from the Terminal window.

We're going to start you off with some SimpleCV basics using image files, and then you'll work your way up to reading images from the camera. Once you've got images coming in from the camera, it will be time to try some face detection:

1. Create a new directory within your home directory called *simplecv-test*.

2. Open the web browser and search for an image that interests you. We used a photograph of raspberries from Wikipedia and renamed it *raspberries.jpg*.

3. Right-click on the image and click "Save Image As."

4. Save the image within the *simplecv-test* folder.

5. In the File Manager (on the Accessories menu), open the *simplecv-test* folder and right-click in the folder. Choose Create New→Blank File.

6. Name the file *image-display.py*.

7. Double-click on the newly created *image-display.py* file to open it in Leafpad.

8. Enter the code in Example 9-1.

9. Save the *image-display.py* file and run it from the Terminal window. If you've got everything right, you'll see a photo in a new window as in Figure 9-8. You can close the window itself, or in the Terminal, press Ctrl-C to end the script.

Figure 9-8. *The raspberry photo displayed in a window*

Example 9-1. Source code for image-display.py

```
from SimpleCV import Image, Display  ❶
from time import sleep

myDisplay = Display()  ❷

raspberryImage = Image("raspberries.jpg")  ❸

raspberryImage.save(myDisplay)  ❹

while not myDisplay.isDone():  ❺
    sleep(0.1)
```

❶   Import SimpleCV's image and display functions.

❷   Creates a new window object.

❸   Loads the image file *raspberries.jpg* into memory as the object **image**.

❹   Display the image in the window.

❺   Prevent the script from ending immediately after displaying the image.

Modifying an Image

Now that you can load an image into memory and display it on the screen, the next step is to modify the image before displaying it (doing this does not modify the image file itself; it simply modifies the copy of the image that's held in memory):

1. Save the *image-display.py* file as *superimpose.py*.
2. Make the enhancements to the code that are shown in Example 9-2.
3. Save the file and run it from the command line.
4. You should see the same image, but now superimposed with the shape and the text.

Example 9-2. Source code for superimpose.py

```python
from SimpleCV import Image, Display, DrawingLayer, Color  ❶
from time import sleep

myDisplay = Display()

raspberryImage = Image("rapberries.jpg")

myDrawingLayer = DrawingLayer((raspberryImage.width,
    raspberryImage.height))  ❷

myDrawingLayer.rectangle((50, 20), (250, 60), filled=True)  ❸
myDrawingLayer.setFontSize(45)
myDrawingLayer.text("Raspberries!", (50, 20),
    color=Color.WHITE)  ❹

raspberryImage.addDrawingLayer(myDrawingLayer)  ❺
raspberryImage.applyLayers()  ❻

raspberryImage.save(myDisplay)

while not myDisplay.isDone():
    sleep(0.1)
```

❶ Import SimpleCV's drawing layer and color functions, in addition to the image and display functions you imported previously.

❷ Create a new drawing layer that's the same size as the image.

❸ On the layer, draw a rectangle from the coordinates 50, 20 to 250, 60 and make it filled.

❹ On the layer, write the text "Raspberries!" at 50, 20 in the color white.

❺ Add myDrawingLayer to raspberryImage.

❻ Merge the layers that have been added into raspberryImage (Figure 9-9 shows the new image).

Figure 9-9. *The modified raspberry photo*

Instead of displaying the image on screen, if you wanted to simply save your modifications to a file, Example 9-3 shows how the code would look.

Example 9-3. Source code for superimpose-save.py

```
from SimpleCV import Image, DrawingLayer, Color
from time import sleep

raspberryImage = Image("raspberries.jpg")

myDrawingLayer = DrawingLayer((raspberryImage.width,
  raspberryImage.height))

myDrawingLayer.rectangle((50, 20), (250, 60), filled=True)
myDrawingLayer.setFontSize(45)
myDrawingLayer.text("Raspberries!", (50, 20), color=Color.WHITE)

raspberryImage.addDrawingLayer(myDrawingLayer)
raspberryImage.applyLayers()

raspberryImage.save("raspberries-titled.jpg")   ❶
```

❶ Save the modified image in memory to a new file called *raspberries-titled.jpg*.

Because this code doesn't even open up a window, you can use it from the command line without the desktop environment running. You could even modify the code to watermark batches of images with a single command.

And you're not limited to text and rectangles. Here are a few of the other drawing functions available to you with SimpleCV (their full documentation is available from SimpleCV (*http:// bit.ly/2OVsDvH*)):

- Circle
- Ellipse
- Line
- Polygon
- Bezier curve

Accessing the Camera

Luckily, getting a camera's video stream into SimpleCV isn't much different than accessing image files and loading them into memory. To try it out, you can make your own basic camera viewer:

1. Create a new file named *basic-camera.py* and save the code shown in Example 9-4 in it.
2. With your camera plugged in, run the script. You should see a window pop up with a view from the camera, as in Figure 9-10.
3. To close the window, press Ctrl-C in the Terminal.

Figure 9-10. *Outputting camera input to the display*

Example 9-4. Source code for basic-camera.py

```
from SimpleCV import Camera, Display
from time import sleep

myCamera = Camera(prop_set={'width': 320, 'height': 240})  ❶

myDisplay = Display(resolution=(320, 240))  ❷

while not myDisplay.isDone():  ❸
    myCamera.getImage().save(myDisplay)  ❹
    sleep(.1)  ❺
```

❶ Create a new camera object and set the height and width of the image to 320 × 240 for better performance.

❷ Set the size of the window to be 320 × 240 as well.

❸ Loop the indented code below until the window is closed.

❹ Get a frame from the camera and display it in the window.

❺ Wait one-tenth of a second between each frame.

You can even combine the code from the last two examples to make a Python script that will take a picture from the camera and save it as a *.jpg* file:

```
from SimpleCV import Camera
from time import sleep
```

```
myCamera = Camera(prop_set={'width': 320, 'height': 240})

frame = myCamera.getImage()
frame.save("camera-output.jpg")
```

Face Detection

One of the most powerful functions that comes with SimpleCV is called `findHaarFeatures`. It's an algorithm that lets you search within an image for patterns that match a particular profile, or *cascade*. There are a few cascades included with SimpleCV, such as face, nose, eye, mouth, and full body. Alternatively, you can download or generate your own cascade file if need be. `find HaarFeatures` analyzes an image for matches, and if it finds at least one, the function returns the location of those matches within the image. This means that you can detect objects like cars, animals, or people within an image file or from the camera. To try out `findHaarFeatures`, you can do some basic face detection:

1. Create a new file in the *simplecv-test* directory called *face-detector.py*.

2. Enter the code shown in Example 9-5.

3. With your camera plugged in and pointed at a face, run the script from the command line.

4. In the Terminal window, you'll see the location of the faces that `findHaarFeatures` finds. Try moving around and watching these numbers change. Try holding up a photo of a face to the camera to see what happens.

Example 9-5. Source code for face-detector.py

```
from SimpleCV import Camera, Display
from time import sleep

myCamera = Camera(prop_set={'width':320, 'height': 240})

myDisplay = Display(resolution=(320, 240))
```

```
while not myDisplay.isDone():
    frame = myCamera.getImage()
    faces = frame.findHaarFeatures('face.xml')  ❶
    if faces:  ❷
        for face in faces:  ❸
            print "Face at: " + str(face.coordinates())
    else:
        print "No faces detected."
    frame.save(myDisplay)
    sleep(.1)
```

❶ Look for faces in the image frame and save them into a faces object.

❷ If findHaarFatures detected at least one face, execute the indented code below.

❸ For each face in faces, execute the code below (the print statement) with face as an individual face.

If your mug is on screen but you still get the message "No faces detected," try a few troubleshooting steps:

- Do you have enough light? If you're in a dark room, it may be hard for the algorithm to detect your face. Try adding more light.

- This particular Haar cascade is meant to find faces that are in their normal orientation. If you tilt your head too much or the camera isn't level, this will affect the algorithm's ability to find faces.

Project: Raspberry Pi Photobooth

You can combine different libraries together to make Python a powerful tool to do some fairly complex projects. With the GPIO library you learned about in Chapter 7 and SimpleCV, you can make your own Raspberry Pi–based photobooth that's sure to be a big hit at your next party (see Figure 9-11). And with the findHaarFeatures function in SimpleCV, you can enhance your photobooth with a special extra feature: the ability to automatically superimpose fun virtual props like hats, monocles, beards, and mustaches on the people in the photo booth. The code in this project is based on the Mustacheinator project in *Practical*

Computer Vision with SimpleCV by Kurt Demaagd, Anthony Oliver, Nathan Oostendorp, and Katherine Scott (O'Reilly).

Figure 9-11. *Output of the Raspberry Pi Photobooth*

Here's what you'll need to turn your Raspberry Pi into a photo-booth:

- A USB webcam or Raspberry Pi Camera Module
- A monitor
- A pushbutton, any kind you like
- Hookup wire, cut to size
- 1 resistor, 10K ohm

Before you get started, make sure that both the RPi.GPIO and SimpleCV Python libraries are installed and working properly on your Raspberry Pi. See "Testing GPIO in Python" on page 99 and "Installing and Testing SimpleCV" on page 137 for more details:

1. As you did in Chapter 7, connect pin 24 to the pushbutton. One side of the button should be connected to 3.3V, the other to pin 24. Don't forget to use a 10K pull-down resistor between ground and the side of the switch that connects to pin 24.

2. Find or create a small image of a black mustache on a white background and save it as *mustache.png* in a new folder on your Raspberry Pi. You can also download our premade

mustache file in the *ch09* subdirectory of the downloads (see "How to Contact Us" on page 15).

3. In that folder, create a new file called *photobooth.py*, type in the code listed in Example 9-6, and save the file.

Example 9-6. Source code for photobooth.py

```python
from time import sleep, time
from SimpleCV import Camera, Image, Display
import RPi.GPIO as GPIO

myCamera = Camera(prop_set={'width': 320, 'height': 240})
myDisplay = Display(resolution=(320, 240))
stache = Image("mustache.png")
stacheMask = stache.createBinaryMask(color1=(0,0,0),
    color2=(254,254,254))  ❶
stacheMask = stacheMask.invert()  ❷

GPIO.setmode(GPIO.BCM)
GPIO.setup(24, GPIO.IN)

def mustachify(frame):  ❸
    faces = frame.findHaarFeatures('face.xml')
    if faces:
        for face in faces:
            print "Face at: " + str(face.coordinates())
            myFace = face.crop()  ❹
            noses = myFace.findHaarFeatures('nose.xml')
            if noses:
                nose = noses.sortArea()[-1]  ❺
                print "Nose at: " + str(nose.coordinates())
                xmust = face.points[0][0] + nose.x -
                    (stache.width/2)  ❻
                ymust = face.points[0][1] + nose.y +
                    (stache.height/3)  ❼
            else:
                return frame  ❽
            frame = frame.blit(stache, pos=(xmust, ymust),
                mask=stacheMask)  ❾
        return frame  ❿
    else:
        return frame  ⓫

while not myDisplay.isDone():
```

```
inputValue = GPIO.input(24)
frame = myCamera.getImage()
if inputValue == True:
    frame = mustachify(frame)    ⑫
    frame.save("mustache-" + str(time()) + ".jpg")    ⑬
    frame = frame.flipHorizontal()    ⑭
    frame.show()
    sleep(3)    ⑮
else:
    frame = frame.flipHorizontal()    ⑯
    frame.save(myDisplay)
sleep(.05)
```

❶ Create a mask of the mustache, selecting all but black to be transparent (the two parameters, `color1` and `color2`, are the range of colors as RGB values from 0 to 255).

❷ Invert the mask so that only the black pixels in the image are displayed and all other pixels are transparent.

❸ Create a function that takes in a frame and outputs a frame with a superimposed mustache if it can find the face and nose.

❹ Create a subimage of the face so that searching for a nose is quicker.

❺ If there are multiple nose candidates on the face, choose the largest one.

❻ Set the *x* coordinates of the mustache.

❼ Set the *y* coordinates of the mustache.

❽ If no nose is found, just return the frame.

❾ Use the `blit` function (short for BLock Image Transfer) to superimpose the mustache on the frame.

❿ Return the "mustachified" frame.

⓫ If no face is found, just return the frame.

⑫ Pass the frame into the `mustachify` function.

⑬ Save the frame as a JPEG with the current time in the filename.

⑭ Before showing the image, flip it horizontally so that it's a mirror image of the subject.

⑮ Hold the saved image on screen for three seconds.

⑯ If the button isn't pressed, simply flip the live image and display it.

Now you're ready to give it a try. Make sure your camera is connected. Next, go to the terminal, and change to the directory where the mustache illustration and *photobooth.py* are, and then run the script:

```
$ sudo photobooth.py
```

The output of the camera will appear on screen. When the button is pressed, it will identify any faces, add a mustache, and save the image. (All the images will be saved in the same folder with the script.)

Going Further

Practical Computer Vision with SimpleCV

This book by Kurt Demaagd, Anthony Oliver, Nathan Oostendorp, and Katherine Scott is a comprehensive guide to using SimpleCV. It includes plenty of example code and projects to learn more about working with images and computer vision in Python.

10/Python and the Internet

Python has a very active community of developers who often share their work in the form of open source libraries that simplify complex tasks. Some of these libraries make it relatively easy for us to connect our projects to the Internet to do things like get data about the weather, send an email or text message, follow trends on Twitter, or act as a web server.

In this chapter, we're going to take a look at a few ways to create Internet-connected projects with the Raspberry Pi. We'll start off by showing you how to fetch data from the Internet and then move into how you can create your own Raspberry Pi–based web server.

Download Data from a Web Server

When you type an address into your web browser and hit Enter, your browser is the *client*. It establishes a connection with the *server*, which responds with a web page. Of course, a client doesn't have to be a web browser; it can also be a mail application, a weather widget on your phone or computer, or a game that uploads your high score to a global leaderboard. In the first part of this chapter, we're going to focus on projects that use the Raspberry Pi to act as a client. The code you'll be using will connect to Internet servers to get information. Before you can do that, you'll need to install a popular Python library called

Requests that is used for connecting to web servers via *hypertext transfer protocol*, or HTTP.

In order to use Requests in Python, you first need to import it. Within Terminal:

```
$ python
Python 2.7.3rc2 (default, May  6 2012, 20:02:25)
[GCC 4.6.3] on linux2
Type "help", "copyright", "credits" or "license" for more
  information.
>>> import requests
>>>
```

If you don't get any kind of error message, you'll know Requests has been imported in this Python session.

Now you can try it out:

```
>>> r = requests.get('http://www.google.com/')
>>>
```

You may be a bit disappointed at first because it seems like nothing happened. But actually, all the data from the request has been stored in the object r. Here's how you can display the status code:

```
>>> r.status_code
200
```

The HTTP status code 200 means that the request succeeded. There are a few other HTTP status codes in Table 10-1.

Table 10-1. *Common HTTP status codes*

Code	Meaning
200	OK
301	Moved permanently
307	Moved temporarily
401	Unauthorized
404	Not found
500	Server error

If you want to see the contents of the *response* (what the server sends back to you), try the following:

```
>>> r.text
```

If everything worked correctly, what will follow is a large block of text; you may notice some human-readable bits in there, but most of it will be hard to understand. This is the HTML of Google's landing page, which is meant to be interpreted and rendered on screen by a web browser.

However, not all HTTP requests are meant to be rendered by a web browser. Sometimes only data is transmitted, with no information about how it should be displayed. Many sites make these data protocols available to the public so that we can use them to fetch data from and send data to their servers without using a web browser. Such a data protocol specification is commonly called an *application programming interface*, or API. APIs let different pieces of software talk to each other and are popular for sending data from one site to another over the Internet.

For example, let's say you want to make a project that will sit by your door and remind you to take your umbrella with you when rain is expected that day. Instead of setting up your own weather station and figuring out how to forecast the precipitation, you can get the day's forecast from one of many weather APIs out there.

Fetching the Weather Forecast

In order to determine whether or not it will rain today, we'll show you how to use the API from Weather Underground (*http:// www.wunderground.com*).

To use the API, take the following steps:

1. In a web browser, go to Weather Underground's API homepage (*http://www.wunderground.com/weather/api*) and enter your information to sign up.
2. After you've logged in to your account, click Key Settings.
3. Notice that there's a long string of characters in the API key field (Figure 10-1). This is a unique identifier for your account that you'll need to provide in each request. If you abuse its

servers by sending too many requests, it can simply shut off this API key and refuse to complete any further requests until you pay for a higher tier of service.

4. Click Documentation and then Forecast so you can see what kind of data you get when you make a forecast request. At the bottom of the page is an example URL for getting the forecast for San Francisco, CA. Notice that in the URL is your API key:

```
http://api.wunderground.com/api/YourAPIkey/forecast/q/CA/
San_Francisco.json
```

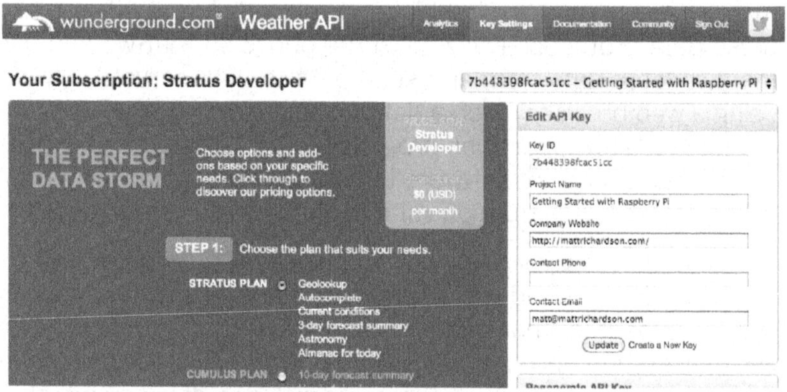

Figure 10-1. *Matt's Weather Underground API account, with the API key in the upper right*

5. To make sure it works, go to this URL into your web browser, and you should see the weather forecast data in a format called *JSON*, or JavaScript Object Notation (see Example 10-1). Notice how the data is structured hierarchically.

Even though the J stands for JavaScript, JSON is used in many programming languages, especially for communicating between applications via an API.

6. To start off, try to get today's local text forecast from within this data structure. Change the URL so that it matches your state and town and put it in a new Python script called *text-forecast.py* (Example 10-2).

7. As you see from the code, in order to get the text forecast for today, we'll have to fetch the right field from the hierarchy of the data structure (see Example 10-1). The text forecast can be found in forecast→txt_forecast→forecastday→0 (the first entry, which is today's forecast)→fcttext.

8. When you run the script from the command line, the output should be your local weather forecast for today, such as this: "Clear. High of 47F. Winds from the NE at 5 to 10 mph."

✏️ Keep in mind that not all APIs are created equal and you'll have to review their documentation to determine if it's the right one for your project. Also, most APIs limit the number of requests you can make, and some even charge you to use their services. Many times, the API providers have a free tier for a small amount of daily requests, which is perfect for experimentation and personal use.

Example 10-1. Partial JSON response from Weather Underground's API

```
"response": {
  "version": "0.1",
  "termsofService":
    "http://www.wunderground.com/weather/api/d/terms.html",
  "features": {
    "forecast": 1
  }
},
"forecast":{  ❶
  "txt_forecast": {  ❷
    "date":"10:00 AM EST",
    "forecastday": [  ❸
      {
```

```
"period":0,
"icon":"partlycloudy",
"icon_url":"http://icons-ak.wxug.com/i/c/k/
  partlycloudy.gif",
"title":"Tuesday",
"fcttext":
  "Partly cloudy. High of 48F. Winds from the NNE at 5 to
    10 mph.",  ❹
"fcttext_metric":
  "Partly cloudy. High of 9C. Winds from the NNE at 10 to
    15 km/h.",
"pop":"0"
},
```

❶ Forecast is the top-level parent of the data we want to access.

❷ Within forecast, we want the text forecast.

❸ Within the text forecast dataset, we want the daily forecasts.

❹ fcttext, the text of the day's forecast.

Example 10-2. Source code for text-forecast.py

```python
import requests

key = 'YOUR KEY HERE'  ❶
ApiUrl = 'http://api.wunderground.com/api/' + key +
  '/forecast/q/NY/New_York.json'

r = requests.get(ApiUrl)  ❷
forecast = r.json()  ❸
print forecast['forecast']['txt_forecast']['forecastday'][0]
  ['fcttext']  ❹
```

❶ Replace this with your API key.

❷ Get the New York City forecast from Weather Underground (replace with your own state and city).

❸ Take the text of the JSON response and parse it into a Python dictionary object.

❹ "Reach into" the data hierarchy to get today's forecast text.

So you now have a Python script that will output the text forecast whenever you need it. But how can your Raspberry Pi determine whether or not it's supposed to rain today? On one hand, you could parse the forecast text to search for words like "rain," "drizzle," "thunderstorms," "showers," and so on, but there's actually a better way. One of the fields in the data from Weather Underground API is labeled **pop**, which stands for *probability of precipitation*. With values ranging from 0 to 100%, it will give you a sense of how likely rain or snow will be.

For a rain forecast indicator, let's say that any probability of precipitation value over 30% is a day that we want to have an umbrella handy:

1. Connect an LED to pin 25, as you did in Figure 6-5.
2. Create a new file called *umbrella-indicator.py* and use the code in Example 10-3. Don't forget to put in your own API key and the location in the Weather Underground API URL.
3. Run the script as root with the command **sudo python umbrella-indicator.py**.

Example 10-3. Source code for umbrella-indicator.py

```python
import requests
import RPi.GPIO as GPIO
import time

GPIO.setmode(GPIO.BCM)
GPIO.setup(25, GPIO.OUT)

key = 'YOUR KEY HERE'   ❶
ApiUrl = 'http://api.wunderground.com/api/' + key +
  '/forecast/q/NY/New_York.json'

while True:
    r = requests.get(ApiUrl)
    forecast = r.json()
    popValue = forecast['forecast']['txt_forecast']['forecastday']
      [0]['pop']   ❷
    popValue = int(popValue)   ❸
```

```
if popValue >= 30:  ❹
    GPIO.output(25, GPIO.HIGH)
else:  ❺
    GPIO.output(25, GPIO.LOW)

time.sleep(180) # 3 minutes  ❻
```

❶   As before, change this to your API key.

❷   Get today's probability of precipitation and store it in pop
     Value.

❸   Convert popValue from a string into an integer so that we
     can evaluate it as a number.

❹   If the value is greater than 30, then turn the LED on.

❺   Otherwise, turn the LED off.

❻   Wait three minutes before checking again so that the script
     stays within the API limit of 500 requests per day.

Press Ctrl-C to quit the program when you're done.

The Weather Underground API is one of a plethora of different
APIs that you can experiment with. Table 10-2 lists a few other
sites and services that have APIs.

Table 10-2. *Popular application programming interfaces*

Site	API Reference URL
Facebook	https://developers.facebook.com
Flickr	http://www.flickr.com/services/api
Foursquare	https://developer.foursquare.com
Reddit	https://www.reddit.com/dev/api
Twilio	http://www.twilio.com
Twitter	https://dev.twitter.com
YouTube	https://developers.google.com/youtube

Serving Pi (Be a Web Server)

Not only can you use the Raspberry Pi to get data from servers via the Internet, but your Pi can also act as a server itself. There are many different web servers that you can install on the Raspberry Pi. Traditional web servers, like Apache or lighttpd, serve the files from your board to clients. Most of the time, servers like these are sending HTML files and images to make web pages, but they can also serve sound, video, executable programs, and much more.

However, there's a new breed of tools that extend programming languages like Python, Ruby, and JavaScript to create web servers that dynamically generate the HTML when they receive HTTP requests from a web browser. This is a great way to trigger physical events, store data, or check the value of a sensor remotely via a web browser. You can even create your own JSON API for an electronics project!

Flask Basics

We're going to use a Python web framework called Flask (*http://flask.pocoo.org*) to turn the Raspberry Pi into a dynamic web server. While there's a lot you can do with Flask "out of the box," it also supports many different extensions for doing things, such as user authentication, generating forms, and using databases. You also have access to the wide variety of standard Python libraries that are available to you.

Here's how to install Flask and its dependencies:

```
$ sudo pip install flask
```

To test the installation, create a new file called *hello-flask.py* with the code from Example 10-4. Don't worry if it looks a bit overwhelming at first; you don't need to understand what every line of code means right now. The block of code that's most important is the one that contains the string "Hello World!"

Example 10-4. Source code for hello-flask.py

```python
from flask import Flask
app = Flask(__name__)  ❶

@app.route("/")  ❷
def hello():
    return "Hello World!"  ❸

if __name__ == "__main__":  ❹
    app.run(host='0.0.0.0', port=80, debug=True)  ❺
```

❶ Create a Flask object called app.

❷ Run the code below when someone accesses the root URL of the server.

❸ Send the text "Hello World!" to the client.

❹ If this script was run directly from the command line.

❺ Have the server listen on port 80 and report any errors.

Before you run the script, you need to know your Raspberry Pi's IP address (see "The Network" on page 40). An alternative is to install avahi-daemon (run sudo apt-get install avahi-daemon from the command line). This lets you access the Pi on your local network through the address *http://raspberrypi.local*. If you're accessing the Raspberry Pi web server from a Windows machine, you may need to put Bonjour Services (*http://bit.ly/1sjViwr*) on it for this to work.

Now you're ready to run the server, which you'll have to do as root:

```
$ sudo python hello-flask.py
 * Running on http://0.0.0.0:80/ (Press CTRL+C to quit)
 * Restarting with stat
 * Debugger is active!
 * Debugger pin code: 238-627-191
```

From another computer on the same network as the Raspberry Pi, type your Raspberry Pi's IP address into a web browser. If your browser displays "Hello World!", you know you've got it configured correctly. You may also notice that a few lines appear in the terminal of the Raspberry Pi:

```
10.0.1.100 - - [19/Nov/2012 00:31:31] "GET / HTTP/1.1" 200 -
10.0.1.100 - - [19/Nov/2012 00:31:31] "GET /favicon.ico HTTP/
    1.1" 404 -
```

The first line shows that the web browser requested the root URL and our server returned HTTP status code 200 for "OK." The second line is a request that many web browsers send automatically to get a small icon called a *favicon* to display next to the URL in the browser's address bar. Our server doesn't have a *favicon.ico* file, so it returned HTTP status code 404 to indicate that the URL was not found.

If you want to send the browser a site formatted in proper HTML, it doesn't make a lot of sense to put all the HTML into your Python script. Flask uses a template engine called Jinja2 (*http://jinja.pocoo.org/docs/templates*) so that you can use separate HTML files with placeholders for spots where you want dynamic data to be inserted.

If you've still got *hello-flask.py* running, press Ctrl-C to kill it.

To make a template, create a new file called *hello-template.py* with the code from Example 10-5. In the same directory with *hello-template.py*, create a subdirectory called *templates*. In the *templates* subdirectory, create a file called *main.html* and insert the code from Example 10-6. Anything in double curly braces within the HTML template is interpreted as a variable that would be passed to it from the Python script via the `render_template` function.

Example 10-5. Source code for hello-template.py

```python
from flask import Flask, render_template
import datetime
app = Flask(__name__)

@app.route("/")
def hello():
    now = datetime.datetime.now()  ❶
    timeString = now.strftime("%Y-%m-%d %H:%M")  ❷
    templateData = {
        'title' : 'HELLO!',
        'time': timeString
        }  ❸
    return render_template('main.html', **templateData)  ❹

if __name__ == "__main__":
    app.run(host='0.0.0.0', port=80, debug=True)
```

❶ Get the current time and store it in now.

❷ Create a formatted string using the date and time from the now object.

❸ Create a *dictionary* of variables (a set of *keys*, such as title, that are associated with values, such as HELLO!) to pass into the template.

❹ Return the *main.html* template to the web browser using the variables in the templateData dictionary.

Example 10-6. Source code for templates/main.html

```html
<!DOCTYPE html>
    <head>
        <title>{{ title }}</title>  ❶
    </head>

    <body>
        <h1>Hello, World!</h1>
        <h2>The date and time on the server is: {{ time }}</h2>  ❷
```

```
  </body>
</html>
```

❶   Use the `title` variable in the HTML title of the site.

❷   Use the `time` variable on the page.

Now, when you run *hello-template.py* (as before, you need to use sudo to run it) and pull up your Raspberry Pi's address in your web browser, you should see a formatted HTML page with the title "HELLO!" and the Raspberry Pi's current date and time.

While it's dependent on how your network is set up, it's unlikely that this page is accessible from outside your local network via the Internet. If you'd like to make the page available from outside your local network, you'll need to configure your router for port forwarding. Refer to your router's documentation for more information about how to do this.

Connecting the Web to the Real World

You can use Flask with other Python libraries to bring additional functionality to your site. For example, with the RPi.GPIO Python module (see Chapter 7), you can create a website that interfaces with the physical world. To try it out, hook up three buttons or switches to pins 23, 24, and 25 in the same way as the Simple Soundboard project in Figure 7-2.

The following code expands the functionality of *hello-template.py*, so copy it to a new file called *hello-gpio.py*. Add the RPi.GPIO module and a new *route* for reading the buttons, as we've done in Example 10-7. The new route will take a variable from the requested URL and use that to determine which pin to read.

You'll also need to create a new template called *pin.html* (it's not very different from *main.html*, so you may want to copy *main.html* to *pin.html* and make the appropriate changes, as in Example 10-8):

Example 10-7. Modified source code for hello-pgio.py

```python
from flask import Flask, render_template
import datetime
import RPi.GPIO as GPIO
app = Flask(__name__)

GPIO.setmode(GPIO.BCM)

@app.route("/")
def hello():
    now = datetime.datetime.now()
    timeString = now.strftime("%Y-%m-%d %H:%M")
    templateData = {
        'title' : 'HELLO!',
        'time': timeString
        }
    return render_template('main.html', **templateData)

@app.route("/readPin/<pin>")    ❶
def readPin(pin):
    try:    ❷
        GPIO.setup(int(pin), GPIO.IN)    ❸
        if GPIO.input(int(pin)) == True:    ❹
            response = "Pin number " + pin + " is high!"
        else:    ❺
            response = "Pin number " + pin + " is low!"
    except:    ❻
        response = "There was an error reading pin " + pin + "."

    templateData = {
        'title' : 'Status of Pin' + pin,
        'response' : response
        }

    return render_template('pin.html', **templateData)

if __name__ == "__main__":
    app.run(host='0.0.0.0', port=80, debug=True)
```

❶　Add a dynamic route with pin number as a variable.

❷　If the code indented below raises an exception, run the code in the except block.

❸ Take the pin number from the URL, convert it into an integer, and set it as an input.

❹ If the pin is high, set the response text to say that it's high.

❺ Otherwise, set the response text to say that it's low.

❻ If there was an error reading the pin, set the response to indicate that.

Example 10-8. Source code for templates/ pin.html

```
<!DOCTYPE html>
  <head>
    <title>{{ title }}</title>  ❶
  </head>

  <body>
    <h1>Pin Status</h1>
    <h2>{{ response }}</h2>  ❷
  </body>
</html>
```

❶ Insert the title provided from *hello-gpio.py* into the page's title.

❷ Place the response from *hello-gpio.py* on the page inside HTML heading tags.

With this script running, when you point your web browser to your Raspberry Pi's IP address, you should see the standard "Hello World!" page we created before. But add */readPin/24* to the end of the URL so that it looks something like *http:// 10.0.1.103/readPin/24*. A page should display showing that the pin is being read as low. Now hold down the button connected to pin 24 and refresh the page; it should now show up as high!

Try the other buttons as well by changing the URL. The great part about this code is that we only had to write the function to read the pin once and create the HTML page once, but it's almost as though there are separate web pages for each of the pins!

Project: WebLamp

In Chapter 6, we showed you how to use Raspberry Pi as a simple AC outlet timer in "Project: Cron Lamp Timer" on page 92. Now that you know how to use Python and Flask, you can now control the state of a lamp over the Web. This basic project is simply a starting point for creating Internet-connected devices with the Raspberry Pi.

And just like how the previous Flask example showed how you can have the same code work on multiple pins, you'll set up this project so that if you want to control more devices in the future, it's easy to add:

1. The hardware setup for this project is exactly the same as the "Project: Cron Lamp Timer" on page 92, so all the parts you need are listed there.

2. Connect the PowerSwitch Tail II relay to pin 25, just as you did in the Cron Lamp Timer project.

3. If you have another PowerSwitch Tail II relay, connect it to pin 24 to control a second AC device. Otherwise, just connect an LED to pin 24. We're simply using it to demonstrate how multiple devices can be controlled with the same code.

4. Create a new directory in your home directory called *WebLamp*.

5. In *WebLamp*, create a file called *weblamp.py* and put in the code from Example 10-9.

6. Create a new directory within *WebLamp* called *templates*.

7. Inside *templates*, create the file *main.html*. The source code of this file can be found in Example 10-10.

In terminal, navigate to the *WebLamp* directory and start the server (be sure to use Ctrl-C to kill any other Flask server you have running first):

```
pi@raspberrypi ~/WebLamp $ sudo python weblamp.py
```

Open your mobile phone's web browser and enter your Raspberry Pi's IP address in the address bar, as shown in Figure 10-2.

Device Listing and Status

The coffee maker is currently on (<u>turn off</u>)

The lamp is currently off (<u>turn on</u>)

Figure 10-2. *The device interface, as viewed through a mobile browser*

Example 10-9. Source code for weblamp.py

```python
import RPi.GPIO as GPIO
from flask import Flask, render_template, request
app = Flask(__name__)

GPIO.setmode(GPIO.BCM)

pins = {
    24 : {'name' : 'coffee maker', 'state' : GPIO.LOW},
    25 : {'name' : 'lamp', 'state' : GPIO.LOW}
    } ❶

for pin in pins: ❷
    GPIO.setup(pin, GPIO.OUT)
    GPIO.output(pin, GPIO.LOW)

@app.route("/")
def main():
```

```python
    for pin in pins:
        pins[pin]['state'] = GPIO.input(pin)    ❸
    templateData = {
        'pins' : pins    ❹
        }
    return render_template('main.html', **templateData)    ❺

@app.route("/<changePin>/<action>")    ❻
def action(changePin, action):
    changePin = int(changePin)    ❼
    deviceName = pins[changePin]['name']    ❽
    if action == "on":    ❾
        GPIO.output(changePin, GPIO.HIGH)    ❿
        message = "Turned " + deviceName + " on."    ⓫
    if action == "off":
        GPIO.output(changePin, GPIO.LOW)
        message = "Turned " + deviceName + " off."
    if action == "toggle":
        GPIO.output(changePin, not GPIO.input(changePin))    ⓬
        message = "Toggled " + deviceName + "."

    for pin in pins:
        pins[pin]['state'] = GPIO.input(pin)    ⓭

    templateData = {
        'message' : message,
        'pins' : pins
    }    ⓮

    return render_template('main.html', **templateData)

if __name__ == "__main__":
    app.run(host='0.0.0.0', port=80, debug=True)
```

❶   Create a dictionary called **pins** to store the pin number, name, and pin state.

❷   Set each pin as an output and make it low.

❸   For each pin, read the pin state and store it in the **pins** dictionary.

❹   Put the **pins** dictionary into the template data dictionary.

❺   Pass the template data into the template *main.html* and return it to the user.

❻ The function below is executed when someone requests a URL with the pin number and action in it.

❼ Convert the pin from the URL into an integer.

❽ Get the device name for the pin being changed.

❾ If the action part of the URL is "on," execute the code indented below.

❿ Set the pin high.

⓫ Save the status message to be passed into the template.

⓬ Read the pin and set it to whatever it isn't (i.e., toggle it).

⓭ For each pin, read the pin state and store it in the pins dictionary.

⓮ Along with the pins dictionary, put the message into the template data dictionary.

Example 10-10. Source code for templates/main.html

```html
<!DOCTYPE html>
<head>
    <title>Current Status</title>
</head>

<body>
    <h1>Device Listing and Status</h1>

    {% for pin in pins %}  ❶
    <p>The {{ pins[pin].name }}  ❷
    {% if pins[pin].state == true %}  ❸
        is currently on (<a href="/{{pin}}/off">turn off</a>)
    {% else %}  ❹
        is currently off (<a href="/{{pin}}/on">turn on</a>)
    {% endif %}
    </p>
    {% endfor %}

    {% if message %}  ❺
    <h2>{{ message }}</h2>
    {% endif %}
```

```
</body>
</html>
```

❶ Run through each pin in the pins dictionary.

❷ Print the name of the pin.

❸ If the pin is high, print that the device is on and link to URL to turn it off.

❹ Otherwise, print that the device is off and link to the URL to turn it on.

❺ If a message was passed into the template, print it.

The best part about writing the code in this way is that you can very easily add as many devices that the hardware will support. Simply add the information about the device to the `pins` dictionary. When you restart the server, the new device will appear in the status list and its control URLs will work automatically.

There's another great feature built in: if you want to be able to flip the switch on a device with a single tap on your phone, you can create a bookmark to the address *http://ipaddress/pin/toggle*. That URL will check the pin's current state and switch it.

Going Further

Requests (http://docs.python-requests.org/en/latest)
   The homepage for Requests includes very comprehensive documentation complete with easy-to-understand examples.

Flask (http://flask.pocoo.org)
   There's a lot more to Flask that we didn't cover. The official site outlines Flask's full feature set.

Flask Extensions (http://flask.pocoo.org/extensions)
   Flask extensions make it easy to add functionality to your site.

A/Writing an SD Card Image

While this book has focused on the Raspbian operating system, there are many other distributions that can run on the Raspberry Pi. With any of them, you need to simply download the disk image and then not-so-simply copy the disk image onto the SD card. Here's how to create an SD card from a disk image on OS X, Windows, and Linux.

Writing an SD Card from OS X

1. Open your Terminal utility (it's in */Applications/Utilities*) to get a command-line prompt.

2. *Without* the card in your computer's SD card reader, type **df -h**. The df program shows your free space, but it also shows which disk volumes are mounted.

3. Now insert the SD card and run **df -h** again.

4. Look at the list of mounted volumes and determine which one is the SD card by comparing it to the previous output. For example, an SD card mounts on our computer as */Volumes/Untitled*, and the device name is */dev/disk3s1*. Depending on the configuration of your computer, this name may vary. Names are assigned as devices are mounted, so you may see a higher number if you have other devices or disk images mounted in the Finder. Write the card's device name down.

5. To write to the card, you'll have to unmount it first. Unmount it by typing **sudo diskutil unmount /dev/disk3s1** (using the device name you got from the previous step instead of **/dev/disk3s1**). Note that you *must* use the command line or Disk Utility to unmount. If you just eject it from the Finder, you'll have to take it out and reinsert it (and you'll still need to unmount it from the command line or Disk Utility). If the card fails to unmount, make sure to close any Finder windows that might be open on the card.

6. Next, you'll need to figure out the raw device name of the card. Take your device name and replace **disk** with **rdisk**, and leave off the s1 (which is the partition number). For example, the raw device name for the device **/dev/disk3s1** is **/dev/rdisk3**.

It is really important that you get the raw device name correct! You can overwrite your hard drive and lose data if you start writing to your hard drive instead of the SD card. Use **df** again to double-check before you continue.

---

7. Make sure that the downloaded image is unzipped and sitting in your home directory. You'll be using the Unix utility **dd** to copy the image bit by bit to the SD card. The following code snippet is the command; just replace the name of the disk image with the one you downloaded, and replace **/dev/rdisk3** with the raw device name of the SD card from step 6.

   You can learn more about the command line in Chapter 3, but you're essentially telling **dd** to run as root and copy the input file (**if**) to the output file (**of**):

   ```
   sudo dd bs=1m if=~/2012-09-18-wheezy-raspbian.img of=/dev/
      rdisk3
   ```

8. It will take a few minutes to copy the whole disk image. Unfortunately, **dd** does not provide any visual feedback, so you'll just have to wait. When it's done, it will show you some

statistics; eject the SD card and you're ready to try it on the Pi.

Writing an SD Card from Windows

1. Download the Win32DiskImager (*http://bit.ly/1riTcHG*) program.

2. Insert the SD card in your reader and note the drive letter that pops up in Windows Explorer.

3. Open Win32DiskImager and select the Raspbian disk image.

4. Select the SD card's drive letter, then click Write. If Win32DiskImager has problems writing to the card, try reformatting it in Windows Explorer.

5. Eject the SD card and put it in your Raspberry Pi; you're good to go!

Writing an SD Card from Linux

The instructions for Linux are similar to those for the Mac:

1. Open a new shell and *without* the card in the reader, type **df -h** to see which disk volumes are mounted.

2. Now insert the SD card and run **df -h** again.

3. Look at the list of mounted volumes and determine which one is the SD card by comparing it to the previous output. Find the device name, which should be something like **/dev/sdd1**. Depending on the configuration of your computer, this name may vary. Write the card's device name down.

4. To write to the card, you'll have to unmount it first. Unmount it by typing **umount /dev/sdd1** (using the device name you got from the previous step instead of **/dev/sdd1**). If the card fails to unmount, make sure it is not the working directory in any open shells.

5. Next, you'll need to figure out the raw device name of the card, which is the device name without the partition number.

For example, if your device name was /dev/sdd1, the raw device name is /dev/sdd.

 It is really important that you get the raw device name correct! You can overwrite your hard drive and lose data if you start writing to your hard drive instead of the SD card. Use df again to double-check before you continue.

6. Make sure that the downloaded image is unzipped and sitting in your home directory. You'll be using the Unix utility **dd** to copy the image bit by bit to the SD card. Here is the command—just replace the name of the disk image with the one you downloaded, and replace /dev/sdd with the raw device name of the SD card from step 5:

```
sudo dd bs=1M if=~/2012-09-18-wheezy-raspbian.img of=/dev/
    sdd
```

This command tells **dd** to run as root and copy the input file (if) to the output file (of).

7. It will take a few minutes to copy the whole disk image. Unfortunately, **dd** does not provide any visual feedback, so you'll just have to wait.

When it's done, it will show you some statistics. It should be OK to eject the disk, but just to make sure it is safe, run **sudo sync**, which will flush the filesystem write buffers.

8. Eject the card and insert it in your Raspberry Pi. Good to go!

BerryBoot

A second way to get the OS onto an SD card is to use the Berry-Boot utility (*http://bit.ly/1BWa6Cg*). BerryBoot is part of the BerryTerminal thin client project, and will let you put multiple operating systems on a single card. You put the BerryBoot image on an SD card, boot it up on the Raspberry Pi, and an interactive installer allows you to select an OS from a list. Note that you'll have to be connected to a network for BerryBoot to work.

Index

Symbols

(hash) tags and cron, 96
--help option, 34
-a switch, 33
. (dot), 32
.. (dots), 32
802.11 WiFi USB dongles, 10
| (pipe) operator, 34

A

absolute paths, 32
Adafruit Industries
    ADS1115 breakout board for, 118
    online store, 82
Adafruit online store, 2
Adafruit: MCP230xx GPIO
    Expander on the Raspberry Pi
    (webpage), 98
ADS1115 (Texas Instruments),
    118-124
Advanced Linux Sound Architec-
    ture (ALSA), 43
AlaMode shield (WyoLum), 76
alsamixer program, 43
Amiga, vii
analog input/output, 113-127
    for Arduino, 74
    converting from digital, 114
    converting to digital, 118-124
    variable resistors, 124
analog-to-digital converter (ADC),
    118-124
Apache web server, 159
API (application programming
    interface), 153, 158
apt-get command-line utility, 42
Arcade Game Coffee Table project,
    xiii

Arch Linux, 15, 46
Arduino, 67-77
    communicating with, 71-75
    default font, changing, 71
    environment location, 69
    Firmatta and, 75
    installing, 69
    powering, 69
    PWM on, 117
    Python in, 54
    on serial port, finding, 72
    serial protocols, 75
    tutorials webpage, 68
    user experience, improving, 71
assembly code, xi
associative arrays (Python), 63
Atom feeds, grabbing, 62
Atrix lapdock, 12
audio out (Raspberry Pi), 5
    forcing output to, 43
    omxplayer, 28
    sending sound to, 110
autocomplete, 29
avahi-daemon, 160

B

Banzi, Massimo, 68
bare-metal computer hacking, xi
Barrett, Daniel J, 44
bash terminal shell, 29
Basic programming language, 48
BBC Micro, vii
BBC News, viii
Behringer's U-Control devices, 11
Berdahl, Edgar, 48
BerryBoot, 175
BitTorrent, 16
blit function, 149
Bonjour Services, 160

About the Authors

Matt Richardson is a San Francisco–based product evangelist for Raspberry Pi, and is responsible for outreach within the United States. He's a graduate of New York University's Interactive Telecommunications Program. Highlights from his work include the Descriptive Camera (a camera that outputs a text description instead of a photo) and The Enough Already (a DIY celebrity-silencing device). Matt's work has been featured at The Nevada Museum of Art, The Rome International Photography Festival, and Milan Design Week, and has garnered attention from *The New York Times*, *Wired*, and *New York Magazine*.

Shawn Wallace is the director of AS220 Industries, part of the AS220 community arts center in Providence, RI. There he shepherds the Providence Fab Lab, Printshop, and Media Arts programming, designs open hardware kits for Modern Device, and runs the local node of the Fab Academy. He's also the member of the Fluxama artist collective responsible for new iOS musical instruments such as Noisemusick and Doctor Om. Shawn was formerly an editor at O'Reilly and Maker Media and is a cofounder of the SMT Computing Society.

Colophon

The cover image is by Marc De Vinck. The cover and body font is Benton Sans, the heading font is Serifa, and the code font is Bitstream Vera Sans Mono.

CPSIA information can be obtained
at www.ICGtesting.com
Printed in the USA
FSOW03n0357120716
22530FS